A Voyage of Discovery, into the South Sea and Beering's Straits, for the Purpose of Exploring a North-East Passage

Undertaken in the Years 1815–1818

VOLUME 2

OTTO VON KOTZEBUE
TRANSLATED BY H.E. LLOYD

CAMBRIDGE
UNIVERSITY PRESS

CAMBRIDGE UNIVERSITY PRESS

Cambridge, New York, Melbourne, Madrid, Cape Town,
Singapore, São Paolo, Delhi, Mexico City

Published in the United States of America by Cambridge University Press, New York

www.cambridge.org
Information on this title: www.cambridge.org/9781108057585

This edition first published 1821
This digitally printed version 2013

ISBN 978-1-108-05758-5 Paperback

The original edition of this book contains a number of colour plates, which have been reproduced in black and white. Colour versions of these images can be found online at www.cambridge.org/9781108057585

CAMBRIDGE LIBRARY COLLECTION

Books of enduring scholarly value

Maritime Exploration

This series includes accounts, by eye-witnesses and contemporaries, of voyages by Europeans to the Americas, Asia, Australasia and the Pacific during the colonial period. Driven by the military and commercial interests of powers including Britain, France and the Netherlands, particularly the East India Companies, these expeditions brought back a wealth of information on climate, natural resources, topography, and distant civilisations. Their detailed observations provide fascinating historical data for climatologists, ecologists and anthropologists, and the accounts of the mariners' experiences on their long and dangerous voyages are full of human interest.

A Voyage of Discovery, into the South Sea and Beering's Straits, for the Purpose of Exploring a North-East Passage

An admirer of Captain Cook, Otto von Kotzebue (1787–1846) was a leading navigator, in Russian service, circumnavigating the globe three times. His 1815 expedition set out to find a passage through the Arctic, study the coastlines of Kamchatka and Alaska, and explore the Pacific. Among the personnel were the naturalist Chamisso and the artist Choris, who both contributed valuable information to the published account, while Eschscholtz, a physician, collected zoological specimens. Originally published in 1821 in Russian and German, this English translation, presented with many plates and charts, appeared the same year and formed part of Darwin's library aboard the *Beagle*. Volume 2 contains the concluding part of the journal, together with contributions by the other members of the expedition. Chamisso's chapters on the Pacific languages are still of interest today, with his glossaries including words not given in any other contemporary sources.

Cambridge University Press has long been a pioneer in the reissuing of out-of-print titles from its own backlist, producing digital reprints of books that are still sought after by scholars and students but could not be reprinted economically using traditional technology. The Cambridge Library Collection extends this activity to a wider range of books which are still of importance to researchers and professionals, either for the source material they contain, or as landmarks in the history of their academic discipline.

Drawing from the world-renowned collections in the Cambridge University Library and other partner libraries, and guided by the advice of experts in each subject area, Cambridge University Press is using state-of-the-art scanning machines in its own Printing House to capture the content of each book selected for inclusion. The files are processed to give a consistently clear, crisp image, and the books finished to the high quality standard for which the Press is recognised around the world. The latest print-on-demand technology ensures that the books will remain available indefinitely, and that orders for single or multiple copies can quickly be supplied.

The Cambridge Library Collection brings back to life books of enduring scholarly value (including out-of-copyright works originally issued by other publishers) across a wide range of disciplines in the humanities and social sciences and in science and technology.

KOTZEBUE'S

VOYAGE OF DISCOVERY.

VOL. II.

London:
Printed by A. & R. Spottiswoode,
New-Street-Square.

I. Clark sc.

Rarick.

London, Published by Longman, Hurst, Rees, Orme & Brown, 1821.

A

VOYAGE OF DISCOVERY,

INTO THE

SOUTH SEA

AND

BEERING'S STRAITS,

FOR THE PURPOSE OF

EXPLORING A NORTH-EAST PASSAGE,

UNDERTAKEN

IN THE YEARS 1815—1818,

AT THE EXPENSE OF HIS HIGHNESS THE CHANCELLOR OF THE EMPIRE,

COUNT ROMANZOFF,

IN THE SHIP RURICK,

UNDER THE COMMAND OF THE LIEUTENANT IN THE

RUSSIAN IMPERIAL NAVY,

OTTO VON KOTZEBUE.

ILLUSTRATED WITH NUMEROUS PLATES AND MAPS.

IN THREE VOLUMES.

VOL. II.

LONDON:

PRINTED FOR

LONGMAN, HURST, REES, ORME, AND BROWN,

PATERNOSTER-ROW.

1821.

A

VOYAGE OF DISCOVERY.

CHAP. XI.

FROM THE SANDWICH ISLANDS TO RADACK.

THE 17th of December, latitude 19° 44′, longitude 160° 7′. Since we left the island of Woahoo, we have always had either calm, or a very faint wind from S.E.; besides this, the strong current from S.W. has carried us in three days 45 miles to N.E.; but it has now taken its direction to S.W.

On the 21st, at six o'clock in the evening, we were in latitude 16° 55′, longitude 169° 16′, consequently on the same parallel, and 15 miles distant from Cornwallis Island. A sailor sat constantly at the mast-head without descrying land, which, however, we could not doubt to be near at hand, on account of the great number of sea-fowl which hovered round us. As soon as the sun sunk into the ocean, I kept the ship against the wind, and tacked during the night under few sails, in hopes of descrying the island, which must be very low, unless its situation had been determined quite inaccurately.

The 22d, at break of day, I again directed my course to W., and expected every moment to hear it announced from the mast-head that land had been discovered; but in vain! At eight o'clock, according to the ship's reckoning, we were in latitude 16° 56′, longitude 169° 21′, consequently almost on the point on which the island was said to lie, without discovering any thing. The many sea-fowls, however, still kept up my hope; but at noon, when our longitude far exceeded that of the island, I was obliged to give it up, as it now appeared probable that the Cornwallis frigate had been mistaken in fixing the longitude. Our latitude, according to observation, at noon, was 17° 3′, longitude 170° 1′; we had consequently been carried by the current six miles to the N. in 24 hours, and must certainly, if their latitude had been accurately determined, have sailed past Cornwallis islands at such a short distance, that we must have seen them had they been ever so low. At noon I gave up looking further for the islands, being convinced that I had sailed by them, which was confirmed by the small number of sea-fowl, and now steered to S.W., with the intention of coming into the latitude of the island of St. Pedro, the longitude of which I wished to determine, if it existed at all. A strong E. wind favoured our course; the weather was serene, but a continual fog seemed to rest on the horizon: this weather, so unusual in the tropics, operated on the barometer, which stood a line lower than usual, which I had never previously observed in the tropics. We had a violent wind on the 24th of December, when we were in latitude 14° 42′, longitude 173° 10′, and which lasted till the 27th, when we were in latitude 11° 3′, and longitude 179° 28′. From the 26th to the 28th we had sailed on the parallel of the island of St. Pedro, 2° from E. to W., without descrying the island, from which it is to be supposed that it does not exist at all, or must lie somewhere else. Here I took a southern course to reach the parallel 10°, which I intended to follow directly W. Since we left the neighbourhood of the Cornwallis islands, we daily saw sea-fowls, and sometimes in such numbers, that there certainly must be here undiscovered islands; though a sailor was constantly on the mast-head, from the rising to the setting of the sun, I was not fortunate enough to make any new discoveries. It might perhaps be supposed, that our longitude was not correct; against which I have to allege daily observation of the sun and moon, from the 23d to the 29th, which would certainly have shewed me the errors of my chronometers.

The 29th, latitude 9° 52′, longitude 186° 34′. That we might be certain not to sail past the chain of Mulgrave islands, near which we ought to be now, I tacked during night under few sails, and continued our course in the morning towards the W.

The 30th, latitude 9° 48′, longitude 187° 9′. We had a fresh wind from N., and extremely high waves from N.W. The current had carried us since yesterday 27 miles to W.: we saw more birds than usual.

The 31st, latitude 9° 49′ 57″, longitude, according to the ship's reckoning, deduced from the chronometers, 188° 33′. During the whole day we had gloomy weather with drizzling rain, which is not usual in the tropics; the wind varied from N. to N.N.E. and N.N.W. This changeable weather gave me great hopes that land was near; I however looked for it in vain. At three o'clock in the afternoon, we were exactly N., 15 miles distant from our course in the preceding year, and had now sailed, for the second time, through the chain of the Mulgrave islands, without having seen them. I now directed my course as much north as the wind permitted, to sail to Kutusof's islands, which I have promised to examine more closely; and we tacked during the night under few sails, that we might neither run aground, nor pass the islands.

The 1st of January, 1817, latitude 10° 10′, longitude, calculated according to the chronometers on the 2d of January, 189° 54′. A faint wind, which changed from N.N.W. to N.N.E., with drizzling rain. At four o'clock in the afternoon, the weather cleared up, and it was announced from the mast-head that land was seen to the N.N.W., exactly in the direction of our course. It was a low, woody island, the length of which, from N. to S., is three miles, and its breadth three-quarters of a mile. As no single islands are known in this part, I concluded it to be a new discovery, and called it New Year's Island, it having been first seen on new-year's day. The faint wind did not permit us to make any further examination this day; a countless number of fish sported round the Rurick: we saw fewer birds, from which I concluded the island was inhabited. We tacked during the night; the weather was uncommonly fine; the full moon looked magnificently in the starry heavens, and guarded us against every danger.

The 2d, at break of day, the island lay distant from us five miles, W. by N. As on the north side a very long reef extended to the north, I directed my course to the south, where we saw no surf, and where we hoped to be able to effect a landing. We had very fine weather, but only a faint breeze from E.N.E. The lovely verdure of the island had a very pleasing look, and the rising columns of smoke had in our eyes something very inviting. When we were distant but two miles from the south point, we were surprised by seven canoes, each rowed by five or six men, which came directly up to us. Their construction was the same as we had observed the preceding year in the Kutusof's group; but they were considerably smaller, and entirely made of very small boards. This way of patching the boats together, proved a want of timber; they have this inconvenience, that the water which runs in is obliged to be constantly baled out; and as they can only leave the island in a perfect calm, they have neither masts nor sails. After much difficulty, they approached us within 100 fathoms, and then almost lay on their oars to admire the ship with attention. Their behaviour was sensible; we observed neither the cries nor grotesque movements with which savages generally distinguish themselves on their first meeting with Europeans: their whole attention was engaged with the ship, which they surveyed with astonishment, from the mast-head down to the water. We did not look upon these savages with less curiosity, who all appeared to me tall and well-shaped; the naturally dark colour of their bodies appeared black at a small distance, because they are tatooed; they do not tatoo their faces. A high forehead, an aquiline nose, and sparkling brown eyes, advantageously distinguish the natives of New Year's Island from the rest of the South Sea islanders. They rub their long black hair with cocoa oil, tie it together above the forehead, and adorn it with flowers and wreaths made of shells; round their neck they wear ornaments of red shells. Their dress was different; some had a couple of fine mats tied round the body; others wore a braided belt, from which the fibres of grass hung down to the feet, and covered them entirely. We were most struck by their ear-holes, which measure more than three inches in diameter, in which they wore green leaves rolled together: some had placed in them such rolls of tortoise-shell. Each boat had a commander, who was easily recognised, as he did not row, and merely gave orders. He is always seated on one side of the boat, with his legs under him, on a raised seat, where he appears with much decorum. One of these chiefs, a tall, handsome man, with a thick beard, was more tatooed than the rest; he held in his hand a large shell, from which he often produced a loud and hollow sound. I do not know what was his motive for doing this; but I remember to have observed such shells in the Marquesas islands, which were never used but in war.

After repeated signs of invitation, and shewing them some iron, they consented to approach the ships, but none of them would venture on board. An animated barter began; for small pieces of old iron hoops, they readily gave the most laborious and skilful of their works; and the leader even gave his beautiful shell horn for a piece of old iron, which he concealed in his belt, after having looked at it with rapture. They dealt with great honesty; and their character seemed to me lively and witty. Their miserable arms, which were carelessly-made lances, prove that they are no warriors. Their other utensils are more elegant than I have seen elsewhere, and they observed the utmost cleanliness in their persons. The island seems to produce but very little fruit and other provisions; at least they had nothing with them except a few grains of pandanus, which they constantly chewed. As far as we could trust the casual view which we had taken of the inhabitants of the Kutusof's islands, they seemed to belong to the same race.

As no breeze agitated the sea, we were enabled to make very excellent observations. The S.W. point of the island lay distant from us three miles to the north, according to the true compass; we found the latitude of the middle of the island 10° 8′ 27″ N.; the longitude, corrected after the latest lunar distance, 189° 4′ 46″ W. I took advantage of the calm, and sent Lieutenant Schischmareff, accompanied by our scientific gentlemen, in two armed boats, to effect a landing if possible. They returned in a few hours, without being able to accomplish their object. The observations of Lieutenant Schischmareff on his attempt, resemble those made by D'Entre Casteaux on Vendola, one of the Admiralty islands, and are as follows:

REPORT OF LIEUTENANT SCHISCHMAREFF.

When the islanders, who were in the boats near the ship, saw that we left them and rowed towards the island, they immediately followed us; we approached a place which seemed convenient for landing; the surf was not violent, and we might have landed, had not the islanders prevented us. A great number of them had assembled on the shore, armed with lances, on which barbed heads were fastened; others surrounded our boats with theirs, and I therefore determined to begin a barter with them on the water near to the shore: they threw themselves in crowds into the water, swam up to us, and brought us mats, necklaces made of shells, cocoa nuts, pandanus fruit, and fresh water in cocoa shells; they also offered us their lances in exchange, as well as two small bows of boards, which they had made into a weapon by means of sharks' teeth, they were a foot and a half in length, and about two inches broad. The number of islanders swimming about continued to increase, and the boats had formed a complete circle round us; but we would not suffer any of them to come too near to ours. They became bold, and even impudent, and offered us cocoa shells filled with sea-water: an old man would absolutely get into the boat in which I was sitting. I struck him on the hands, threatened him with my sabre;

but he would not desist from his purpose, till I gave him a blow with the fist on his head; upon which he left me, and swam on shore. Another old man wanted to take possession of the rudder of the Baydare, at which the pilot was so enraged that he was going to shoot him, had he not been prevented by the gentlemen on board of her; and to avoid disagreeable consequences, I determined to return to the ship. The island is surrounded with red coral reefs: at the place in which we were, the water was not more than a foot in depth. At the reef the depth is five fathoms; and about fifteen fathoms from the shore, we could not reach the bottom; the islanders had assembled on this reef, on which account we could not have landed without danger. There were about eighteen boats which surrounded us, none held more than six men; several only one or two; and they were all without masts. The number of islanders in the boats, and on the shore, amounted to about two hundred; we saw but a few women among them, and no children. The island is very woody, and we saw a great many pandanus, but only a very few and low cocoa trees. The islanders were all tatooed, like those that came to the ship, and likewise had rolled leaves in their ears.

That we had begun the new year with a discovery, seemed to us a good omen, and gave us all much pleasure. I now changed my plan of sailing to the Kutusof's group, and, as it might be conjectured that there were several other groups near New Year's Island, I steered, by the aid of a faint N. wind, to the S.W., to come again into the parallel of 10°, and then to take a westerly course. The parallel 10° in longitude, 189° in Arrowsmith's map, is strewed with such a great number of islands that I could not miss them, even if the tenth part of them existed. When the sun set, we lost sight of New Year's Island, though we were distant from it only a few miles, and tacked during the night under few sails, to keep the ship on one point.

The 3d of January.—Latitude 9° 59′ 47″ N., longitude 189° 38′, we continued our course to W., saw a number of snipes, but no land, and were soon, according to the ship's reckoning, in latitude 10° 2′, longitude 189° 40′. I now thought it unnecessary to go further to W., being convinced that the islands marked on Arrowsmith's map are not to be found here, and therefore ordered the ship to tack to S.E., to try my luck in that direction. After we had followed this course till seven o'clock in the evening, we were in latitude 9° 37′, seven miles west of the course of the preceding year, without being able to descry land from the mast-head. Immediately before sun set, we punished a pelican for its temerity in flying so close over our heads that we might have caught it with our hands; a shot brought it down into the sea, and I sent out a boat, notwithstanding the high waves, to fetch this prize for our collection of natural history. During the night we kept our station under a few sails: a serene horizon is very rare in these parts, a thick fog always seems to rest on it.

The 4th. Latitude 9° 43′, longitude 189° 53′. I intended to remain only for to-day in these parts: at break of day I caused all the sails to be set to follow the course to W. by S. till noon, and then tack to S.E. A fresh wind from N.N.E. favoured our course; the Rurick ran seven knots; it was near noon, and our hope had almost sunk, when we suddenly heard the joyful news from the mast-head that land was seen to the S. by W. At one o'clock, we could already see from the quarter-deck, at a distance of six miles, a chain of small woody islands, the spaces between which were filled up by reefs; they extended as far as the eye could reach, and I had now counted above twenty. I followed the chain at a distance of two miles, and saw the surf furiously raging between the spaces, and the water at the other side of the chain quite smooth. At four o'clock in the afternoon we reached the western point of the group; here the islands ended, but a long reef, which appeared a little above the water, extended to S.W., and then took its direction to S.E. further than the eye could reach. As soon as we had sailed round the western point of the reef, we were under the wind in very still water, and as I hoped to be able to discover a passage between the reefs, I approached them within 200 fathoms, and sailed along them at the same distance. I knew from experience that the depth between such coral reefs is always considerable, and I therefore approached nearer, braving the danger which threatened us; besides this, it is the only way to examine them, as, even at the distance of half a mile, the passage is no longer visible. D'Entre Casteaux, who, in surveying the coast of New California, likewise hoped to find a passage between the reefs, says he only approached them within three miles, and therefore could not possibly make the desired discoveries. Such a dangerous navigation always requires the greatest precaution; the sailor must never quit the mast-head, a second must be on the bowsprit, a third on the fore part of the ship (the gin), and the mate, provided with a good telescope, must sit in the round top to give timely warning of every danger. Captain Flinders, with justice, says, in his voyage, when he was examining Torres Strait, which is likewise strewed over with countless reefs, "He who has no strong nerves ought to leave such an attempt alone." Mine allowed me to face the danger, though every sudden change of the wind threatened us with death, because the ship might be immediately wrecked. I and all my companions were on our guard, and the crew was prepared every moment to put the ship about. With these precautions we continued our way very quickly, without missing the least bending or opening of the reef. The chain lay to the north of us, at the distance of six miles; the way thither was saved us by the reef, which was two fathoms broad; upon the other side of it the water was smooth, and the depth appeared considerable. As far as we could see, the reef extended to S.E., and we observed, at the end of it, from the mast-head, a small island lying higher than the rest, and which probably united with the reef. At last, to our great joy, we discovered two passages, through which we hoped to be able to pass with the Rurick, though they were very narrow. This discovery, which is important, not only to us, but to every navigator, would certainly have been passed over by us, had we not approached the reef within musket-shot. It was too late to make any examination to-day, and we left our dangerous situation for the night.

The 5th of January. Latitude 9° 27′ 55″ N., longitude 190° 11′ 30″. The current had carried us so far to N.W. during the night that we could not see any land; we however again got sight of the island at seven o'clock, and at nine were at the same point we left yesterday. I now sent Lieutenant Schischmareff to explore the most northern of the two passages, in which he found a considerable depth, but judged it impossible to force the ship in, as the fair way or channel was seldom more than fifty fathoms broad, made continual turnings, and besides this, the entrance was in such a direction that the monsoon always blew out of it. The other passage, which was four miles more to the south, had now to be explored; we reached it at noon, and were with the Rurick 200 fathoms from its entrance, and, while Schischmareff was sounding the channel, we took observations of the latitude and longitude of this strait. We saw Schischmareff get happily between the reefs, upon which he made the signal that the channel itself had no bottom, but at the place where he then was, 100 fathoms, and at the other side of the reef 26 fathoms water, upon a coral bottom: the narrowest pass in the strait he stated at 123 fathoms. This information gave me uncommon satisfaction, as I could now hope to reach it with the ship, and to make this very remarkable discovery. These islands inspire great interest, merely by their nature, as they owe their origin entirely to marine animals; and I determined to hazard a great deal, before I gave up the plan of penetrating this chain of islands. The boat was called back as a fresh wind sprung up, and rendered further examination difficult; our situation was also dangerous as the day was drawing to a close, and my whole attention was bent to find a means to remain during the night in this interesting, but dangerous spot. At length I hit upon one which, hazardous as it was, we however adopted, not to be obliged to leave the place. It was this: Lieutenant Schischmareff went with a stream anchor to the reef, fastened it to it, and when he had given a signal that he had succeeded, I steered thither with the Rurick, brought it within 50 fathoms of the place, took in all the sails, and then the ship was fastened to the stream anchor with a cable of 175 fathoms in length. The die was cast, and as long as the monsoon continued to blow from N.E. there was no danger, but had it turned to the S.E., which is frequently the case, our fate would have been inevitable. The Rurick now stood in the middle of the ocean, fastened to a coral reef, under the protection of God, in whom I trusted; a strange feeling seized me in this singular situation; a look on the raging sea made me shudder; but when I again turned my eyes towards the channel, the most pleasing hope animated me, to which I willingly resigned myself. The reefs consist principally of dark coral, which is mixed with but a little red; at low water the rocks are visible for about two feet, which was the case when we fixed the stream anchor; but all was soon covered by the rising tide. We had, at a small distance from it, 40 fathoms' depth, but which increased a little further so much, that we could find no bottom. On the east side of the channel a little sandy island has formed itself, but which will certainly extend in time, and when covered with vegetation, will assume the form of the other islands. A number of sharks surrounded us, which devoured every thing thrown overboard; they seemed to keep chiefly about the channel, probably because many fish swim in and out with the regular current. We saw many flying fish rising in the air, I suppose to avoid their enemies. The boat which was sounding the channel was attacked by sharks, which would not be driven away by striking them with the oars; two of them were caught, which was very easy, as they swallowed the hook immediately it was thrown in the water. We had now ended our examinations; a violent gust of wind from E.N.E., which half an hour before would have alarmed us very much, gave us but little uneasiness, as the cable held fast. At midnight we observed the current from the channel, which, when it was most rapid, ran one knot.

The 6th of January, at four o'clock in the morning, when it was still quite dark, the wind changed to E., and soon after to E. by S., which brought the Rurick so close to the reef, that it was easy to throw a stone into the breakers; the depth here was 23 fathoms. As no more than another point to the S. would suffice to dash us against the rocks, I was obliged to quit my post, and even to leave the stream anchor behind, which, being deep under water, would have taken too much time. The cable was loosened, the sails hoisted, and we retired from the reef without injury, tacking near it. As soon as the sun appeared above the horizon, we again directed our course to the entrance, found the wind from E. and E. by S., embraced this moment, as it permitted us to run into the channel without tacking, into which we sailed with all our sails spread. At forty minutes past nine we were in the middle of the channel. A death-like silence prevailed on board the ship; we heard on both sides the roaring of the breakers, and every one was ready at his post. At length, the mate called from the mast-head that we were out of danger, for the water now assumed a dark colour. The Rurick, in fact, was sailing over water as smooth as a mirror; we had the breakers behind us, congratulated each other on the success of our hazardous attempt, and looked eagerly at the islands, which we hoped soon to reach. The current, which run two knots in the narrow part, had rapidly carried us past every danger, for we had been only fifteen minutes from the beginning to the end of the strait. We gave the channel the name of Rurick's Strait. The wind allowed us to steer direct to the westernmost island, where we saw, on the 4th *, from W. to E., columns of smoke, and, by means of a telescope, some people also, a sight which heightened our triumph at having penetrated hither. In spite of our desire to become acquainted with the inhabitants of this

* To avoid mistakes I shall number the islands, counting from W. to E.

unknown island, we sailed slowly and with precaution, and often threw out the lead that we might not run aground. Immediately after we left Rurick's Strait, the depth was from 26 to 27 fathoms, over a ground of living coral; but as we approached the island it regularly decreased, and at a distance of two miles we found it 18 fathoms. The ground, which sometimes consisted of fine coral sand, induced us to hope that we should find a good anchoring place near the island, and Lieutenant Schischmareff, who had gone on before with the boat, soon made us a signal, that he had found ten fathoms' depth over fine coral sand, upon which we directed our course thither.* In the N. lay, at a distance of 200 fathoms, the reef which united the third with the fourth island. At the same distance we were protected to the E. by a coral reef, visible at low water, and we lay in quite smooth water, which was not agitated in this place, even during the most violent wind. Our view was for the present confined in the E. to the chain of islands; in the W. we saw the reef, round the outside of which we had sailed; in the S. we had the clear horizon before us: even from the mast-head, the reef through which we had sailed could not be discovered, and only the small high island which I

* On the chart drawn by us here, our anchorage, and the situation of the islands, are distinctly marked. Vide Plan of the group of islands Romanzoff.

have mentioned before, was visible. The farther geography of this group of islands was still enveloped in darkness; yet it was to be conjectured that there must be a connection to the east, because no high waves came from that side. We were at first so agreeably occupied with our situation, that we postponed the farther examination. I resolved not to quit the place till I had determined it astronomically by various observations, tried and regulated my chronometers, and done some work on board the ship, which could not be done in open sea. At our anchoring place we found the water so transparent, that we could see the bottom at a depth of from 10 to 12 fathoms; at the same time we had the most beautiful weather. Our naturalists, who had made an excursion into the third island, returned in the evening with shells and plants, much satisfied with their success. The islands 1, 2, and 3, we found uninhabited, though we everywhere discovered traces of human beings. At three o'clock in the afternoon, we saw a boat under sail, coming from the E., which stopped at the fourth island, landed something there, and then directed its course to us. From the large sail, and the skilful manœuvring we observed, that it entirely resembled that of the Kutusof's group. It approached the Rurick within 50 fathoms, the sail was taken in, and an old man at the helm, probably the commander, showed us some fruit, speaking a great deal. The many times repeated word, *Aidara*, reminded us of New Year's Island, where we had also heard it. We did not succeed in enticing them nearer to the ship; they always contrived to keep at a certain distance from us by tacking; they looked at the ship with much curiosity, but they took no notice of us. At length I sent off a little boat to them, but as soon as they saw it put off, they retired, and when our people overtook them, they, terrified at their approach, threw into it bread-fruit, cocoa-nuts, and pandanus; some pieces of iron, however, which were shown them lessened their fear, and they joyfully accepted them. They now talked a great deal without our being able to understand one another, and at length the savages left us, taking their course to the fourth island, to which they invited us by signs. From this first meeting it was to be concluded, that we had to deal with a very kind-hearted people. The manner of their tattooing and dressing is the same as on New Year's Island, and it is highly probable that it is the same race. We found ourselves very happy in being in this secure place, as the weather was gloomy, and the wind very high.

Early on the 7th of January, I sent the long-boat and the Baydare for the stream-anchor which we had left behind, and which they safely brought on board in the afternoon. To-day our new acquaintance manœuvred round our ship, all adorned with wreaths of flowers, and in the highest state. I sent Lieutenant Schischmareff and M. Chamisso to the fourth island, to try to conciliate the confidence of the inhabitants, who would on no account approach us; and as soon as the old man perceived that our boat was going that way, he did not stay a moment, but followed her with the loudest expressions of joy. We saw both boats put on shore, and lieutenant Schischmareff will give an account of what happened there.

REPORT OF LIEUTENANT SCHISCHMAREFF.

"In consequence of your command I have been to Goat Island, in order to become acquainted with its inhabitants. I steered directly up to a place which appeared to me convenient to land, and the islanders also stood for it in their boats. On landing, I observed several people walking about their huts, who immediately fled into the woods on seeing us. I landed near a hut, and as I found nobody in it, I did not proceed farther, but waited for the islanders who had not been able to land where I did, but had been obliged to steer to a spot a quarter of a mile lower down. I left my men in the boat, and went up to them alone; six men got out of the boat; some of the fugitives came up to them, but soon withdrew, as well as three of those who had just landed; the other three came up to me. I did not understand what drove them into the woods, whether fear, or a plan for a surprise; but which did not appear dangerous to me, as I had two pistols about me, and had my armed men at hand; but when they approached me, I saw that they were without any arms, and that all three appeared afraid of me. They kept at a distance of twenty paces from me; an elderly man held something white in his hand, on leaves, which he seemed to design for me, but would not venture to come nearer; meantime he broke off the bough of a tree, probably as an emblem of peace: I immediately did the same, and went up to him. The man at first timidly drew back, but he at last offered his present, always repeating the word *Aidara*. I received his present; and, though I did not know the meaning of it, I also said after him, *Aidara*. I have learnt, in the sequel, that *Aidara* means Friend. Upon this, the woman who was with him, and who was probably his wife, gave me a pandanus branch; and the third person, a young man of about twenty years of age, who had no present ready for me, presented me with his own necklace, with which I adorned my hat; the old man then took a wreath of flowers from his head, which I directly put on mine. This, it seems, gave them courage, and we went together to the huts, where we were joined by our naturalist, whom they in the same manner presented with a necklace and wreath. I had now the iron brought, with which I returned their presents; the rest from the woods now came up, and also received iron;

they were much rejoiced at it, and expressed their gratitude and joy by repeated exclamations, and friendly behaviour We were now surrounded by all the islanders, thirteen in number; they shewed themselves open and friendly, but always a little shy: they were all unarmed. The company consisted of a man of about forty years of age, two elderly and one young woman, and three young men of twenty years' old, and some children from nine to fifteen; one of them was but three year's old, and was still in arms; the oldest man had a short black beard, short black hair, and wore a short mat; the others had no beard: the young men also wore mats, but the children were quite naked. The women were wrapped in mats from the hips to the feet; they were all of a pretty dark colour, but slender and slightly built. They all seemed very clean; the men were painted with different squares of a dark blue colour, like those on New Year's Island; the women had not much of this painting on their neck and bosom: they all wore rolled leaves in their ear-holes; the neck was adorned with a row of shells, and they wore wreaths of flowers on their heads. They have a great resemblance with the natives of New Year's Island; their countenance expressed much good nature. I should here observe, that we had one family before us of which the elderly man was the head. Our naturalist gave to the father some seeds of water-melon, and showed him how to plant them; and I enquired, as well as I was able, whence they procured the water which I found with them in cocoa shells. They understood me, and conducted me to a place almost in the middle of the island, where the rain-water collects in a pool from the higher grounds. We now went to the shore, where we met with some large trees which had been drifted there by the sea, and resembled our oaks. Coming up again to the huts, the chief invited us to enter his, which consisted of a roof resting on four poles; two mats were laid down in the inside, on which we were invited to sit. A woman dressed a pandanus fruit, which she bruised with a stone; the man then pressed the juice into a shell; and though all this was done with the hands, it was done with the greatest cleanliness; and as he was presenting us with the juice, and something fell into it, he did not take it out with his fingers, but with a chip. During this time our sailors were entertained in another hut. I made the chief a present of two knives and some pieces of iron, and M. Chamisso of some fishing-hooks. We invited him to come to our ships and he seemed much pleased with us. We had thus formed a new acquaintance, and the word *Aidara* was often exchanged between us. Our new friends also accompanied us to the shore, and assisted us in getting our boat into the water."

The small number of people whom Schischmareff met with, leads me to suppose that they must have their proper abode elsewhere, and only visit this island at times. A white lump was brought me, resembling white chalk, the use of which I did not learn till I had become more acquainted with the islanders. A plant grows here, called by the natives mogomuk, and its root, which resembles a small potatoe, is dried in the sun, and rubbed to powder; it produces a fine flour, which, when pressed into lumps, may be kept a long time without spoiling. When it is to be eaten, a little is broken off one of these lumps, mixed with water in a cocoa-shell, and boiled till it becomes a thick pap: its taste is not amiss, and has a great resemblance to our potatoes: this plant grows wild.

Early on the 8th we saw the boat with our friends sailing to the E., where it soon vanished. They probably went to distant islands to carry the news of the arrival of the large ship with the white men. At noon, I went on shore, in the hope of meeting with some people; but they were all gone; which I regretted the more, as I had provided myself with useful presents; six goats, a cock and a hen, and various kinds of seeds, and yams, were all objects not to be expected here, and with which I hoped to enrich them for the future. We landed opposite the hut where Schischmareff had been yesterday so kindly received. The goats received their liberty, and immediately fell to upon the fine grass, which, after their long voyage, they found growing beside the huts: the cock, with his hen, immediately mounted on the roof of it, and proclaimed, by loud crowing, that he had taken possession of it, while he tenderly divided with his consort a lizard which he had caught. I planted the yams in the neighbourhood of the habitations, and, in a walk which we made into the interior of the island, M. Chamisso sowed in various places the seeds which we had brought with us. After a slight observation, we found that this island, like all the others, consisted of fragments of coral. The animal builds upwards from the bottom of the sea, and dies as soon as it reaches the surface; from this edifice there is then formed, by the constant washing of the sea, a grey calcareous stone, which seems to be the basis in all the islands, and, being gradually covered with sand, becomes a sandy island, which increases in the course of time, and, by the seeds* which the sea washes up, is covered with plants, which, by the falling off of the leaves, at length produce a black fruitful earth. I cannot enter into a detailed explanation of the origin of the Coral islands, they belong to the

* This seed is of such a nature, that it can be drifted about the sea for years together without spoiling, as it lies concealed in a thick shell. It probably comes from the coasts of America, where it is carried by the rivers into the sea, and, by the strong current which, between the tropics, generally go from E. to W. is at last brought to these islands. To be convinced of the possibility of this, we may remember the Japanese ship which was carried by the current from E. to W. in seventeen months from the coast of Japan to California.

department of the naturalist, from whom the reader must expect further information. The island was covered in many places with an impenetrable wood, where the pandanus, which diffuses an agreeable aromatic smell, is the most frequent kind of tree ; the bread-fruit tree, which arrives here to an immense thickness and height, is also very frequent: the time of the fruits seemed to be past. Cocoa trees are very rare; we, however, found some young trees of this species, which had been lately planted. Of animals, we have only seen rats of a middling size, and lizards; the former are so bold, that they run about us without showing the least fear. We did not see any land birds. In a square pit, in a low part of the island, we found clear water, of so pure a taste, that I sent for some every day. When we left the island, which I called Goat's Island, we saw the goats, with the cock and hen, in their old place. The inhabitants will certainly be much surprised at their new guests, but, as Talso left a piece of iron behind, they will, at the same time, be convinced, that we came here only with good intentions.

The evening and the following night we had a violent squall of wind and rain from E. by N.; and the 9th we spent the morning in various works, and in vain expectation of the inhabitants. In the afternoon I sent a boat for water, and at the same time the mate Chramtschenko was charged to draw a base line, and to take angles. The boat returned in the evening with the news, that people had been found there who had probably arrived in the night. The mate told me that they had received him very kindly, obliged him to take some ornaments which he had refused to accept, treated him with pandanus' juice, and, besides, did not let him want for entertainment. He farther observed, that he had seen neither women nor children, but a very old man whom he had not known before. Chramtschenko found the presents we had brought thither, the buck and the goats having taken for their night's lodging a small hut near the principal habitation. The islanders ventured to cast only stolen glances upon these animals, and were ready to run away every time they moved. It may easily be supposed what an impression such a strange animal with horns and a long beard must have made on the savages, and it is not surprising that they run in all directions with loud cries, on an attempt being made to bring one nearer to them. This description of their fright put me in mind of Robinson's Friday, who had likewise been in the greatest alarm on seeing a goat. The mate tried to make them understand that the goats were a present from us, and intended for their support; which they at length seemed to understand, as they often repeated the word *Aidara*, by which they meant to express their gratitude and friendly intentions. The fowls were known to them; they called the cock *kahu*, and the hen *lia-lia-kahu*.

A piece of stuff, which we had yesterday left in the hut, and which was still lying in the same place, gave them indescribable joy when the mate divided it. We considered this great honesty; but, on a nearer acquaintance, found them to be arrant thieves, and that their reserve was only owing to their fear. During our stay here we made several observations, and took a number of distances.

January 10.—After the longitude and latitude of our anchoring-place had been carefully observed, I resolved to follow the chain of islands farther to the east, and, therefore, dispatched Lieutenant Schischmareff, at five o'clock in the morning, with the long-boat, to look for an anchoring-place, which we might reach in one day from the place where we now were; for as the wind here generally blows from the east, and fresh during the day, no great progress can be made in this direction; which, however, could not divert me from my plan. In the morning we had fine weather, but in the afternoon a violent storm arose, which compelled the long-boat to return to the ship. Lieutenant Schischmareff had proceeded seven miles without finding a safe anchorage; though the nature of the bottom and the depth allowed of anchoring in some places, yet there was no where any protection against the waves from the east, which cause the ship to reel, and the cables are exposed to the danger of being cut by the coral rocks. On his way, he remarked several coral banks which lay southwards from the chain of islands; close to the reefs which connect the islands, the bottom consists of fine sand; but opposite the islands, of living corals. When Schischmareff was passing by Goat Island, he saw men on the intermediate reefs, who were taking advantage of the ebb, to pass from one island to another; all the other islands appeared to him to be uninhabited.

January 11. — As by Schischmareff's account, the examination with the long-boat was exposed to difficulties, partly because the storng wind would not permit it to make much progress, partly because the waves continually dashed over it, I resolved, with the first fair weather, to make an attempt with the Rurick itself. The weather today was not favourable to the attempt; I therefore remained at anchor, and went in the afternoon to Goat Island, to make observations with the dipping needle. I saw no men there, but plenty of rats, which often interrupted me in my work, as I was obliged to drive them away. I found one of the goats dead, probably of indigestion from change of food. I have given our anchoring-place the name of "Christmas Harbour," because we passed that holiday there, according to the old style.

The 12th, the wind blew violently from E. by N., and seemed not favourable to my undertaking; however, we were under sail as early as six o'clock in the morning, because I hoped that the weather would become fine at sun-rise, as had often happened. At first we tacked with good success, but this pleasure did not continue long; the wind grew more and more violent, the air so dark that the land was hidden from us; it was necessary at last to reef the top-sails, by which the sheets were often torn, and much time lost. All these inconveniencies we had borne courageously, but when now the sun, by a single beam, enlightened the surrounding objects, the man at the mast-head, and the mate at the top, cried out, "We are surrounded by shallows; we are in the midst of coral reefs." In an instant the ship was turned against the wind; and it was high time, for we were just on the point of running upon a shoal which the cloudy weather had concealed from us; but we had scarcely time to perceive our critical situation, when the sun was again involved in clouds, and we were again in danger. Most of these banks hardly reach the surface of the sea, are of small extent, and rise perpendicularly from the bottom. In clear weather they may be seen at a considerable distance, as every shoal is marked by a dark spot on the water; but in gloomy weather the whole surface of the sea is of a dark colour, and the danger is not perceived till there is scarcely time to avoid it. This was our case; for as soon as the ship was under weigh we perceived a shoal; we tacked, and tacked again; and at last came rain, which enveloped us in darkness, and gusts of wind which again tore the main-top sheets. I now gave up the idea of proceeding farther, and only thought of carrying our ship back safe to our harbour. After innumerable tackings between coral banks and shoals, we, in fact, succeeded in regaining our anchoring-place, but exhausted with fatigue, having passed three hours in truly painful exertions. In this unsuccessful expedition we had penetrated seven miles to the east, and had descried from the mast-head land in that direction, whence we conjectured that we were in the midst of an archipelago of islands. We had bad weather, not only this day, but the 13th, one gust of wind followed another, and some were so violent that I was in fear for the cable: after a heavy torrent of rain it was calmer. In the afternoon I sent a boat to Goat Island to fetch water, which we did daily, in order to spare the stock of water which we had brought from the Sandwich islands. The rainwater which is caught is unfit for keeping, as it becomes putrid in six days. Meantime we saw a boat, coming from the eastward, land at Goat Island, and the subaltern officer informed me that he had found people there, who had received him in a friendly manner; some of the women even endeavouring to entertain him by singing and dancing. We solemnized New Year's day (old style) by repose, and I busied myself with a plan for making an excursion in boats, as soon as the weather should become more favourable.

On the 14th, in the morning, we again saw a

boat from the east, under sail, which landed on Goat Island; it was the second of the kind, and I now confidently hoped that the abode of these savages must be on the eastern part of the group of islands. The mate, whom I had sent on business to the island, returned with the news, that he had found strange islanders; that he had been treated with boiled fish and baked bread-fruit, and likewise entertained by the women with singing and dancing. Every meeting that we had yet had with the natives proved their gentle dispositions. The goats were still objects of terror to them, and one of them had to-day in particular thrown them into great alarm: the mate advancing to the huts, the oldest man held out a nosegay to him, as a sign of peace; the he-goat ran up to his old companion, and as he ran by, snatched the nosegay with such violence from the hand of the terrified savage, that the latter received a blow with the horns; of course he and all his companions ran off with loud cries, and the mate had a great deal of trouble to entice them back to their habitations, after he had driven the goats into the bushes.

The weather was to-day uncommonly fine, and favourable for our enterprize; no time was lost in making every thing ready, and at one o'clock in the afternoon I left the Rurick, accompanied by our scientific gentlemen, Lieutenant Schischmareff, and several sailors, in two well-armed boats. We were, in all, nineteen persons, and had taken provisions for five days. At three in the afternoon we reached the fifth island, where I resolved to stay for the night, and to continue our voyage to the east at day-break. I purposely made so short a way to-day, that my people, who were obliged to row incessantly, might not be too much fatigued; besides, we all wished to satisfy our curiosity here by examining the corals, as well on the island as on the reef. It happened that we arrived just when the tide was at the highest, and were therefore able conveniently to bring the boats into a channel between the fifth and sixth islands; at the ebb they were indeed left dry, so that we had to take care to get them afloat again at high water, twelve hours afterwards. We pitched our tent on a pleasant meadow, under the shade of a pandanus tree; and, while a fire was lighting to make tea, I strolled about the island with my gun, and our naturalists busied themselves with the corals. I had soon traversed the island in all directions, it being only half a mile in circumference. The interior of it consists of large dead blocks of coral, which are covered with a layer of mould not more than two inches deep at the most, whereas on Goat Island it is in many places at least three feet deep. This difference proves how much more recently this little island has arisen; and, on the whole chain, the remark was subsequently confirmed, that the smaller islands, in comparison with the larger ones, were much behind, and the vegetation but poor, because they are destitute of earth, which is produced only in a long series of years by the falling off and rotting of the old leaves.

The spot on which I stood filled me with astonishment, and I adored in silent admiration the omnipotence of God, who had given even to these minute animals the power to construct such a work. My thoughts were confounded when I consider the immense series of years that must elapse, before such an island can rise from the fathomless abyss of the ocean, and become visible on the surface. At a future period they will assume another shape; all the islands will join and form a circular slip of earth, with a pond or lake in the circle; and this form will again change, as these animals continue building, till they reach the surface, and then the water will one day vanish, and only one great island be visible. It is a strange feeling to walk about on a living island, where all below is actively at work. And to what corner of the earth can we penetrate, where human beings are not already to be found? In the remotest regions of the north, amidst mountains of ice, under the burning sun of the equator, nay, even in the middle of the ocean, on islands which have been formed by animals, they are met with!

I visited also the sixth island, to which, at the ebb, I could pass without wetting my feet, and found it exactly like the fifth. On the shores, which are exposed to the open sea, the breakers rage in a dreadful manner, the foaming spray dashes many fathoms into the air, and long blocks of coral, probably rent from the reef during violent storms, and thrown on shore, lay scattered all around; the shore is covered with a quantity of shells of various kinds, and fragments of coral. After looking in vain for a bird worth adding to our collections, I returned to our resting place, where we all assembled cheerfully round the tea-kettle, and heartily rejoiced at being on so remarkable an island, which we had ourselves discovered. Chamisso and Eschscholz returned with a quantity of rare corals and marine animals; their conversation upon them was instructive, and we listened with attention, till the rats and lizards disturbed us by stealing our biscuit. Chamisso and Eschscholz affirmed that these rats and lizards were not different from those of Europe; the question was started, whence did they come? And this problem can only be solved, —by the shipwreck of a vessel on these islands. The same question arose, when our common flies here pestered us incessantly. We saw many crabs, (Muschelkrebse, *cancer Parasiticus,*) which always carry their dwelling about with them. Thus the evening came on, amidst various interesting researches and observations; and our repast, which consisted of English patent meat, and which we found excellent, filled us with gratitude towards the inventor.

I had a fire kept up during the night; and two sentinels, with their guns loaded, were stationed on both sides of our resting-place, to give the alarm by a shot, if necessary; we also slept in our clothes, with our arms by us. Though these precautions were almost too great among so mild a people, I did not like to depart even here, from the rule which I had laid down, never to neglect the greatest caution. We passed the night undisturbed; but I could not sleep for thinking of the discoveries I expected to make. The fire in the wood in the darkness of the night, the calling of the watch, the wild songs of the savages in Goat Island, the hollow roaring of the surf, the entire strangeness of the place in which I was, produced in me a most singular emotion, which kept me awake, and I almost envied my companions, who were all buried in peaceful sleep.

At three o'clock in the morning, according to my calculation, the water must have reached its height; I hastened to our boats, and found, to my vexation, that there was not water sufficient to get them out; which compelled us to wait to-day for the tide by which we landed. We employed ourselves meantime, as the weather was delightful, in examining the corals, and when the water rose, I ordered the long-boat to be loaded. Just as we were ready to set out upon our voyage, the sailor who kept watch, called out that he saw two boats, one of which was very large, coming from the east; they had observed us, and came directly towards the island. I resolved to wait for the boats, ordered the arms to be placed in readiness, and stood myself, with Schischmareff and our naturalists unarmed, and with eager expectation on the shore. Both soon anchored at a short distance, and the dexterity with which they brought their canoes under the wind, and took in the sails, proved to us that they were very experienced seamen, and gave us great pleasure. Their sails consisted of very finely braided mat, and were shaped with so much art, that even the closest side-wind could not fail to catch them. The large boat, on board of which we counted five and twenty men, was thirty feet long, had on the outriggers a small hut, and a number of ropes hung down from the very high mast.

After they had, amidst much noise, finished their manœuvres, four men leaped into the water, and swam towards us. One of them led the way, bearing a large shell-horn; the others followed with cocoa-nuts and pandanus fruit; and those who had remained behind, awaited in silence the success of their embassy, which advanced towards us with much confidence. The leader, with the horn, was advantageously distinguished by his whole appearance: he was a tall, well-made man, of thirty; his black hair, which was elegantly bound together upon his head, was ornamented with a wreath of white flowers, in the form of a crown. In his ear-holes, which were remarkably large, he wore rolls of tortoise-shell, ornamented with flowers; round his neck hung various gay ornaments: he was differently tattooed, and much more than the others, which gave him the appearance of a man in armour: his face, animated by a pair of most expressive eyes, was adorned with whiskers. Astonishment, fear, and curiosity, alternated in his countenance; but, overcoming his feelings, he advanced towards me with a majestic step, and repeatedly uttering the word *Aidara!* presented to me his shell-horn.*

His companions laid the fruit at my feet, looked at us with constrained friendliness, but trembled very much, particularly one of them, who had really convulsive contractions. We tried to inspire the embassy with courage, in which we succeeded pretty well, except with this one; and they seemed much surprised at our friendly behaviour. I had a red cloth spread on the shore, invited the chief to sit down, and sat down myself next to him, while the others, standing, formed a circle round us. He sat with much dignity after the Asiatic fashion, became more and more animated, and asked us many questions, sometimes pointing to the sea, sometimes to the sun, and then to the sky. I at last conceived that he wished to have it explained whether we came from the sky or the

* I have learnt that the leader makes use of this horn in war, and presents it only to his conqueror. They probably took us for supernatural beings, and consequently thought themselves already vanquished.

sea; and when I gave him to understand that I did not comprehend his language, he was vexed with himself, and talked always louder and quicker, while his looks viewed, without intermission, all the new objects around him, but without interrupting his conversation. When he was particularly pleased with any thing, he could not refrain from laying hold of it to inquire after its use; if we could make him comprehend it, he expressed his astonishment by a loud prolonged *O — h!* His companions, who otherwise did not utter a sound, repeated it; and the third echo of the long *O — h!* resounded from the canoes. He cried aloud to them what he had seen, again repeated his *O — h!* and continued his conversation till a new object attracted his attention. Among other things, he laid hold of a tin box, which he examined on all sides with much curiosity, and, when I opened the lid, started back with a loud *O — h!* He immediately announced the wonderful circumstance to the people in the canoes, and when I opened several boxes he was quite beside himself with astonishment, and there was no end of his monotonous exclamation. I shall call this inquisitive man Rarick, for so he was called by his companions, and I was delighted that his name differed from that of our ship in only a single letter. After I had long entertained myself with this amiable Rarick, and he had forced me to accept shell-wreaths, and several other of his elegantly wrought ornaments, I sent to the boats for some knives, scissars, and iron, and scarcely did the latter, which consisted of pieces of old hoops three inches long, strike their view, when they again repeated their astonishment by a loud *O—h!* and the desire to possess this treasure was evident in their ardent looks. *Möll! Möll!* (so they called the iron)* was now re-echoed from mouth to mouth; a terrible cry was set up in the boats, and six men, who could not resist the attractive charm, leaped into the water, and came to us to look at the iron; we could hear nothing but *Möll! Möll!* I presented Rarick with some pieces of iron, a knife, and scissars: he held his treasure with both hands, pressed it closely to his breast, as if he was afraid of losing it, and could scarcely persuade himself to believe that he possessed such an invaluable treasure, which the others devoured with greedy looks; but when their turn came, their envious countenance cleared up; the rejoicing was beyond bounds; they all danced about with their iron as if mad, crying, *Möll! Möll!* without intermission. Their comrades in the boats became uneasy; some ventured to swim to shore, and when they likewise received presents, the dreadful noise be-

* The use of it is known to these islanders; we found in the sequel several pieces among them, which they had probably procured from the beams of a wrecked vessel. Such beams, which appeared to be of American wood, were afterwards seen by ourselves.

gan anew. The bond of friendship was now concluded; the savages became less reserved, joked, and often embraced us. I tried to make Rarick understand that I intended to go to the east, where I supposed his habitation to be. He understood me, and willingly got into my boat. We set off; Rarick sat next to me, and the savages got their canoe under sail with admirable quickness, in order to put about, as their boats are not made for rowing. But as we were obliged to keep at some distance from them, not to get in each others' way, Rarick's courage failed; to be alone with us appeared dangerous to him, and his fear was evident in all his motions, though he tried to conceal it as much as he could. They often called something out to him from the canoes; the conversation between him and his subjects became more animated, and his anxiety increased with every stroke of the oars: we tried in vain to quiet him. Before we were aware of it, he jumped, with all his treasures, into the water, and swam quickly to his canoe, into which he got, and suddenly turned to Goat Island. It was impossible for us to suppose that they would abandon us, after having made them so many presents; but, probably, they had received tidings respecting the bearded, horned animals, at which they wished to look: and, in fact, my mate, who happened to be there at the time, confirmed my supposition; they had looked at the goats with much surprise, had ran away in great terror when they bleated, and had then laughed at each other for their apprehension. Our companions left us at three o'clock, and we hastened to examine the ninth island before sun-set, which we reached at seven o'clock, and where we determined to remain for the night, as my sailors had been much exhausted in rowing against the wind. We were now distant from the ship five miles, and still continued to see in the E. the open sea. As soon as we arrived, we crossed the island in all directions, to discover people, but found only traces of them, and some huts which had just been left. In the middle of the island stood a house, which entirely resembled that on Goat Island, except that it was considerably larger: it had the form of a Chinese temple; a square roof, very neatly made of rushes, pointed at the top, rested on four pillars, five feet above the ground, which protected it against the scorching rays of the sun, and the wind blew cool between the posts: the floor was paved with coral stones. The interior, from the top of the roof down to the posts, was divided by a handsomely worked lattice, in the middle of which there was a square opening, large enough conveniently to admit a man. The rats have certainly given the islanders the idea of building their huts on posts, for I observed that they had their store-room in the inside of the lattice, to which the rats could not get up the smooth pillars. Their sleeping houses are built on the ground, and consist of a roof, with two entrances: the habitations for the day are so large, that from twenty to thirty people have room in them. The house which we visited was filled with utensils of all kinds; fishing-nets, fish-hooks, lines, vessels of cocoa shells, and the like, lay all in confusion. Its situation made it a pleasant retreat, as it stood in the middle of a small grass-plot, surrounded and shaded by bread-fruit trees, which stood so close together that there was no approach to the house but by a narrow footpath.

This island appeared to us to be older than Goat Island, which we concluded, from the very luxuriant vegetation, and the deep mould. Here, too, cocoa trees were very scarce, but we every where saw young trees of this kind, just planted; it seems as if all the islands had been but lately inhabited. After we had, in vain, looked for inhabitants, we took up our quarters on a green spot, near the shore, and the heat of the day made us all long for the evening. The nights here are uncommonly fine, and have the advantage above those in other warm countries, that no dew falls, as the coral islands send forth no exhalations. You may repose at ease under the starry heavens, refreshed and cooled by the mild monsoon, which blows here, as it is not arrested by any high islands. The kettle with the patent meat assembled us to a cheerful meal, and we then lay down on our grass beds. The azure sky was our canopy; Sirius sparkled brightly over our heads, and a fine warm air breathed around us. The sentinels were placed as in the preceding night, and we had only to defend ourselves against the attacks of the rats. I awoke as the day was dawning, and enjoyed the rising of the sun; never did it appear to me to rise more majestically from the ocean, than on this low island: night was still hovering over the deep, a golden mist on the edge of the horizon announced the approach of the "king of day;" in a few seconds, he appeared in all his splendour, and the ocean, reflecting his radiance, afforded a most enchanting prospect.

On the 16th of January, we were already on our way again; at six o'clock in the morning, the faint wind, and the coolness of the morning, permitted us to row briskly. When we reached the ninth island, the nature of the bottom allowed us to anchor, though not so conveniently as in Christmas Harbour. The higher the sun rose, the fresher the wind became, and we did not reach till noon the thirteenth island, having proceeded four miles from our night's quarters. Here we refreshed ourselves by a meal, and my sailors, whom I did not wish to expose to the fatiguing labour in the burning sun, rested for a few hours. The island was only a mile in circumference, and uninhabited; at least we neither discovered huts, nor reservoirs for water. A reef extends to the south, by which a small harbour was formed on the south-west part, which is protected towards the east. We climbed up a tolerably high tree, and saw land to the south-east, by which I was confirmed in my supposition that we were in a circle of islands. To the N. E. 1½ mile distant from us, lay a small island, which appeared higher than any I had hitherto seen. After dinner, the sentinels informed us that they had seen three men, who were coming from the westward along the reef; they were favoured in this walk by the tide; and though the water is so deep in some places, that they are obliged to wade through it, the inhabitants, however, make use of this way, with as much confidence, as we do of our roads. By means of a telescope, I recognized Rarick, and his companions; they soon came up, unarmed, and were very happy to see us again. My friend was again uncommonly talkative; and though I did not understand a word at first, yet his eloquence furnished us at length with a store of expressions, which we wrote down, when we thought that we had understood them; thus we learnt that a man is called *Manuan*, and a woman *Redgini* *; *Tamon* signifies a leader; and such a one was Rarick over the whole group of islands. With inconceivable vivacity, he addressed himself to me, and no body could undertstand him; at length he named his companions, then himself, and as he then looked at me with an air of inquiry, we guessed that he wanted to know my name.

* The *gi* is pronounced as in French.

He was extremely rejoiced when he found that we understood him, called himself by my name, and me by his, endeavouring at the same time to discover whether I was satisfied with it. As I had, on a former occasion, become acquainted with the custom of the South Sea islanders, to exchange names as a token of friendship, I willingly accepted the proposal. I was now called Rarick, and he Totabu, as he could not pronounce my name in any other manner. The companions of Totabu, at the same time exchanged names with our scientific gentlemen, and, from that time, our intercourse was more confidential. Totabu's learned researches now re-commenced; my fowling-piece, the use of which I would not venture to show him, occupied him very much; our clothes were an object of their highest curiosity, and our shoes created the most immoderate laughter, particularly as one of them, in his attempt to walk in them, fell down at his full length. But they were above all surprised, when I pulled off my jacket, and uncovered my arm; they were astonished on perceiving it was white, and did not cease to utter their loud *O—h!* All the valuables which I had yesterday given to Rarick, he carefully carried about him, wrapped up in pandanus leaves; and, that he might sometimes enjoy himself with looking at them, and quickly to cut something with his scissars, would unwrap them, but immediately concealed them again in his girdle. A small looking glass was never out of his hands;

he looked at himself in it without ceasing; and his companions, to whom he sometimes allowed a look in it, seized upon it to catch a glimpse of their own portraits, which afforded them all much amusement. Rarick's boats now anchored at our island; he begged me to accompany him to his home, pointing to the east, and we resolved that he should go on before, and we follow in our boats. It was four o'clock when we embarked, and I directed my course to the high island to the N.E., which I determined to explore, because its height, compared with the others, appeared to us remarkable. We reached it in an hour, but it was a long time before we could find a landing-place, as it was surrounded by reefs, and, to satisfy our curiosity, we were obliged to wade more than knee deep through the water. The island was about the same size as the one we had left, had scarcely any mould, but was covered with immense blocks of coral, which lay irregularly one over the other, and seemed to have been dashed on shore, which may have been done by a violent storm from the north. In spite of the little earth, trees were already growing between the large blocks, which resembled in height and thickness our largest oaks, and had nearly the same appearance. A number of birds, of the sea-swallow species, which had their nests in these trees, on seeing us, set up a most hideous cry.

As there were likewise very good anchoring-places for a ship at this island, I altered my plan of visiting Rarick in his dwelling. I hoped to come with the Rurick to the end of the group; and, to lose no time, immediately returned, having previously given this high island the name of Bird Island. A long reef extended from it to N.E., at the end of which we descried land; we had observed but few coral banks in our tour. Rarick had, in the meantime, got so much a-head with his boats, which sailed very close to the wind, that he was scarcely visible on the horizon; we also spread our sails, and happily reached our ship in the evening with a fair wind.

The 17th we spent in making the necessary preparations for our approaching voyage; in the evening the wind changed, for the first time, during our stay here, to the N.N.E., and we only wished that it would continue so the following day, as it would considerably facilitate our progress to the east.

OBSERVATIONS MADE IN CHRISTMAS HARBOUR.

Latitude of our anchoring place, the mean of daily observations - - - - - -	9° 32′ 36″ N.
Longitude, by solar and lunar distances, the mean of a great	

number of observations made on different days - -	190°	6′	50″	W.
Variation of the magnetic needle	11	00		E.
Dip of the magnetic needle -	17	55		

On the 6th of January, the day we reached our anchoring-place, our chronometers gave the following longitude, corrected by the latest lunar distance.

Barraud's chronometer -	190°	13′	30″	W.
Hardy's chronometer - -	190	6	48	

According to Barraud's chronometer, the longitude of Christmas Island has been determined, which differed only seven miles from the true. I made this correction afterwards in laying down the chart.

The 18th of January the wind blew moderately from N.N.E.; we weighed anchor, and were already under sail at six o'clock in the morning. As the wind allowed us to-day to take a course parallel with the island, we fortunately avoided the coral banks, which had caused us so much terror in our first attempt. In about three hours, when we were near the eleventh island, the wind veered from N.E., and obliged us to tack; the dangerous navigation began at the thirteenth island; we were again surrounded by coral banks, but we had fair weather, and as we were now more accustomed to the sight of these dangerous cliffs, and had besides always timely warning from the mast-head, the constant tacking of the ship was, it is true, very fatiguing, but we were amply rewarded by the prospect of executing our plan, and were already, at noon, opposite Bird Island. The greatest depth which we found on approaching the middle point of the circle, was thirty-one fathoms; the bottom consisted of living coral, small pieces of which were brought up by the lead; near a reef, the depth was from ten to twelve fathoms, and the bottom consisted of fine coral sand. At four o'clock we reached the seventeenth island, which forms the north point of the whole group, and were distant from it three-fourths of a mile, and cast anchor in fifteen fathoms, upon a bottom of fine coral sand.

We lay here as secure as in the finest harbour, as the ship was quite covered from the N. and E.. and the water as smooth as a mirror. We overlooked the whole eastern part of the group, which consisted of small islands, lying very close together, and from the seventh took a direction to the S.E. Our labour was rewarded, for we had attained our object; for it could not be difficult for us now to explore the chain with a S.E. course. The seventeenth island, which is larger than the Goat Island, afforded us a pleasing prospect by its luxuriant vegetation and large trees, among which there were a great number of cocoa trees. We saw many huts; some people were walking on the shore, and seemed to be admiring our ship; some boats were sailing to S.E.; others coming from thence, and it appeared to us that we were now, for the first time, in the inhabited part of the group. A boat under sail came to visit us, and one of Rarick's companions, whom I immediately recognised, presented me with some cocoa nuts, repeatedly exclaiming *Rarick! Totabu! Aidara!* His joy was increased, on my giving him some iron; but neither he nor his companion would venture on board, though we begged them very much. I had a boat got ready, that our naturalists might have an opportunity of examining the island, and the savages immediately sailed off with them. I intended to remain in the anchoring-place for to-morrow, to be able to determine it more exactly by astronomical observations. This place is twenty miles from Christmas Harbour, in a straight line. In the evening, the gentlemen returned, very much pleased at the friendly reception of the islanders, of whom they saw about thirty; an old man, who, to judge from his ornaments, was a chief, had offered M. Chamisso, as a proof of his good intentions, a meal, made of a mixture of pressed pandanus and bread-fruit, which did not taste amiss; the others had collected round the white men, whom they regarded with much astonishment.

When we inquired for Rarick, they pointed to the S. E.; consequently he resided there, and we might hope that we should meet with more inhabitants than we had hitherto done. We learnt that they called the seventeenth island, Ormed, and an island in general, Enns.

The 19th: our friend from the Goat Island arrived here to-day, but did not venture nearer than twenty fathoms from the ship, and, after showing us cocoa-nuts, and speaking a great deal, hastened to the island of Ormed. This man, who had no reason to complain, as he had received many presents, could not conquer his fear. They admired the ship very much at a distance, gesticulated, and spoke with vehemence, and repeated particularly often, *Ellip Oa!* (large boat). I have observed that the inhabitants of this group advantageously distinguish themselves from those of Easter and Penrhyn's Islands, by their mature consideration and reflection, with which the others do not trouble themselves. After I had determined the situation of the place, I went on shore in the afternoon: my friend of the Goat Island had already introduced me as the *Tamon Oa Ellip* (commander of the great boat); and they all ran to the shore to meet me. A very old man, with a venerable countenance, and a long grey beard, whom, from his look, I took for the chief, said, *Aidara!* presented me with some cocoa nuts, and invited us to go to his hut which was just by, where elegant mats were spread out between four posts, in the middle of which I was invited to sit down. The rest of the company, men, and some very pretty women, with children in their arms, formed a circle round us; all looked at me with the greatest attention, and the most solemn silence prevailed; suddenly it was interrupted; they all sprung up with a loud cry, and ran away, as if they had seen an evil genius, all except the old man, who remained trembling by my side. This confusion had been caused by a dog, which we had brought from the coast of Chili, and as he never quitted me, he had jumped unawares into my boat. To come up to me he was obliged to leap over the shoulder of one of the savages, who were seated in a close circle, when this unexpected circumstance gave occasion to the ludicrous scene, which became still more comic, when the dog, who is generally fearful, encouraged by the cowardice of his opponents, forced them, by his barking, to seek refuge in the trees, which they climbed with the agility of monkeys. I succeeded, with great trouble, in convincing the old man of the harmlessness of the animal, upon which he called his subjects together, who gradually ventured to approach, without taking their eyes from the object of their fear, whose least motion caused them to start. As they are not acquainted with any quadruped except the rat, which is by them called *Didirick*, they named the dog *Didirick Ellip*. It was not till I had sent the evil genius into the boat that their countenances brightened up, and the old man presented me with cocoa-nuts and cakes of a very agreeable taste, prepared from pandanus juice, and which they called *Magan.* I had my presents now brought; a large hatchet and two knives delighted the old man very much, as he had never seen such a large piece of iron, and after I had split a piece of wood with it, the often heard *O—h!* re-echoed in the whole circle. As they are principally employed in ship-building, and make their boats without any other tools but coral stones and shells, it may be easily imagined how invaluable they must have considered the hatchet. If I had afforded the men pleasure by giving them knives, I made the women still more happy with beads and looking-glasses; they did not cease admiring the charming things. They at length were easy respecting their treasures, and they all directed their attention to me; but only the old man ventured to touch me. He told his people a great deal, who listened to him with open mouths; I was likewise obliged to uncover my arm here, which they touched, to convince themselves that it was skin, and not any kind of stuff. I observed, for the first time, a certain modesty in the women, to which the other women of the South Sea Islands are total strangers. The men in vain persuaded their wives to touch my arm; they refused it with much decorum. I have likewise observed in the sequel, the native modesty of these women. After they had sufficiently examined every thing, I resolved to treat them with another sight, and held my watch to the old man's ear, who started back with surprise at the ticking of it; they all listened to it, the gold gave them much pleasure, and the motion of the second-hand astonished them very much; but when I made the watch repeat, my conjuration almost terrified them; they retired, and long and seriously discussed this important subject, till I enticed them to me by some presents. It was now their turn to give me presents; the women gave me their elegantly wrought shell-wreaths, which they took off their heads and placed on mine; the men took off their necklaces, laboriously made of red coral, and gave them to me; the old man gave me a pretty large mat, giving me to understand that I might sleep on it; and, at last, the men and women joined in a song which was addressed to me, and probably expressed their gratitude. On a tour which I made through the island, several of them accompanied me, and one went on before to show me the best way. I was unarmed, for I felt myself quite secure among these kind-hearted children of nature, who, to amuse me, would play and dance on before me. This island appeared to me to be older than any we had seen; I saw here pandanus and bread-fruit trees of uncommon height and thickness; the cocoa tree is met with but seldom, and generally only just planted. Near the houses I observed a plant with beautiful blossoms which they cultivate merely to adorn themselves with the flowers; and this trait alone proves that this people are not entirely in so rude a state as the other savage tribes, and I am convinced that sensible Europeans might raise them to a state of real civilization. On passing a cocoa-tree, I observed a stone tied to a branch of it; I asked my companion why it was done, and received for answer, *tabui*, at the same time giving me to understand that the fruit was not allowed to be eaten. The word *tabui* has much resemblance with the *taboo* of the South Sea islanders, and seems to have the same meaning, but I have never heard it since. It would be remarkable to find words here which could prove, by their similitude, that the inhabitants of this group have perhaps come from the east; but of all the words which we have hitherto marked down, not a single one, except this, indicates such a circumstance. On the shore we met with a plain tomb, consisting of a square built of coral stones; it appeared to me that the inhabitants were not allowed to enter it, and I have since learnt, that only the chiefs are buried, but that all the other corpses are thrown into the sea. It was evening; I was obliged to put an end to my walk, and took leave of my friends, who accompanied me to the boat, where they perceived a fowling-piece, the use of which they insisted on knowing. I gave them to understand that it produced a loud noise, but they misunderstood me, and thought that I used it as they do the shell-horn. The old man gave me some cocoa-nuts on parting, and called me his *Aidara*.

On the 20th of January, early, we were under sail; a stormy wind from N. N. E. favoured our S. E. course parallel with the chain of islands.

We found the latitude of our anchoring-place to be - -	9°	33′	16″	N.
Longitude, according to the chronometers, - - -	189	49	2	W.
Variation of the magnetic needle,	12	14		E.

After an hour's brisk sailing, without being detained by any coral banks, we saw an island in the S. E., which exceeded all the others in height. I steered directly up to it; and was more and more convinced that we were in a circle, as we now also perceived land in the S. At nine o'clock, we cast anchor a quarter of a mile from this large island, in eight fathoms' water, over fine sand; and we were in a very fine harbour, of the calmest water. A boat, which left Ormed at the same time with us, sailed, to our great astonishment, as fast as the Rurick. Being uncertain whether this was Rarick's residence or not, as soon as we had cast anchor, I sent M. Chamisso on shore, to gain information respecting it. In an hour, he returned with the news, that Rarick was here, and would immediately visit me on board; that there was nothing to distinguish it as the residence of a chief; every thing was exactly the same as on Ormed island, even the population scanty, which, of men, women, and children together, consisted of sixty persons. In the afternoon, a boat put off from the island, and we soon recognized Rarick, who, while still at a distance, called out to us *Aidara!* He was very richly adorned to day with flowers, and shell-wreaths; he wore various ornaments round his neck, and his body was wrapped in a new mat. His boat approached our ship, into which he got without hesitating, to our great astonishment, and two of his companions, encouraged by his example, followed him. It would be vain for me to attempt to describe the first moment they got on board; they stood as if petrified, their sparkling eyes wandering on all the objects around them: they would not have moved a step had not I taken hold of Rarick by the hand, and led him about. He recovered by degrees from his astonishment, and was as if re-animated; so active, so inquisitive, and childish, as I had never seen him before. He ran from one object to another, felt it with both his hands, asked its use, never waited for an answer, but flew to something else. He did not dwell a moment on one subject; too many things attracted his attention; curiosity and fear alternated in his features. He leaped about on the deck like a madman, sometimes laughed with all his might; full of astonishment, he would sometimes call out *O—h!* and when he was particularly struck with any thing, cried out *Errio! Errio!* (a word which I often heard on such occasions). His companions, likewise, took a lively interest in every thing, but did not venture to express themselves so loudly in the presence of their chief as he did. I had almost frightened away my friends by my own fault; we had still a couple of hogs left, of our whole stock, which I determined to leave on this island. To learn whether they were acquainted with these animals, I had them brought out; this, however, produced a very great confusion, because they entered squeaking terribly. My guests were thrown into the greatest alarm. Rarick threw both arms round me; he trembled in every limb, and cried out louder than the hogs; and I hastened to send them back. The islanders stared around them with bewildered looks, and even my presents did not avail to restore the good humour with which they came on board. I invited Rarick to come into my cabin, but he cautiously sent his companions on before, who listened with visible anxiety, and slowly descended the stairs; but they had scarcely entered, when their astonishment was without bounds; the many shining things gave them inconceiveable pleasure, which they expressed by calling *Errio! Errio!* and covered their faces with both hands. A glimpse into the glass at first terrified them: they contemplated each other in silence, and then again looked into the glass. But when they recognized each other in it, they embraced, made all kind of comical motions, and laughed immoderately. Rarick, who heard it from above, could contain himself no longer. In one leap he was close to us, and his delight was boundless. I was as if surrounded by romping children, though the very grey beard of one of them showed his advanced age. But I have often observed, that, in this people, age does not suppress childish cheerfulness; and some who could scarcely move for the infirmities of age, took a share in every thing with youthful vivacity; and I never saw them out of temper. Their fine climate, and their diet, consisting of fruits, may be the reason of this circumstance, which is so rare among us: and it is probably also owing to their vegetable diet, that this whole people is tall and very slender. Their bones are as delicate as those of women: their hands and feet uncommonly small. They have very little laborious work to do: their only employment is to build boats, which they now cannot do without; these are long and narrow, and lie deep, on which account they can sail against the wind; the sails and cords are made by the women, with great ingenuity, out of cocoa bark. The people are mild and timid, but appear to be sometimes engaged in war, as they have wooden lances which are badly made, provided at the top with barbs or sharks' teeth, which must certainly inflict very bad wounds. After my friends had been sufficiently entertained in the cabin, I accompanied them on deck, where some more islanders had come, and who heard from their countrymen long accounts of what they had seen. I again gave all

Inside of a House in the Radack Islands.

of them presents, and delighted Rarick very much by a red apron, which I tied round his hips; upon which he immediately sent for a number of cocoa-nuts from his canoe. When he was going on shore, he invited me to accompany him in his canoe, which I accepted, while our scientific gentlemen followed us in a boat. Rarick conducted us to his habitation, which was distinguished from the rest by its spaciousness, and treated us with a beverage made from pandanus juice, and which had a sweet and spicy taste. One of our gentlemen affirmed he had seen a piece of iron which had not come from us; and when I repaired to the place, where a canoe was just then building, I really found a piece, four inches long, and two broad, made use of by the builder instead of an axe. I exerted all my skill in pantomime to discover whence they had it. They understood me, and told me, that a large beam had swum there from N. E., round the middle of which there had been an iron band, which they took off, broke into several pieces, and divided among them. The keel for the new boat, which was hollowed out by means of the little piece of iron, with infinite expense of time, was laid, and it will take at least a year before a boat of twenty feet is finished. The keel is generally made of the bread-fruit tree, and they would be glad to make a whole boat of it, if its fruit did not form part of their food; at present they are obliged to content themselves with wood which is drifted from the east, from distant islands, or from the coast of America, and is sometimes very hard to work. As they cannot make long boards with their miserable instruments, they use for the exterior covering of the boats small pieces of wood, which they bind together with strings of cocoa bark. Their boats, at first sight, appear to be old and patched; but they know so well how to stop up all the holes and spaces, that but very little water can penetrate; perhaps their boats will succeed better in future, by the assistance of the hatchet and axe which I gave them, and in the use of which I instructed them. Rarick and several islanders accompanied us in a walk through the island, which is five miles and a half in circumference. There was no want of excellent mould here, which, in many places, even formed small hills. Bread-fruit and pandanus are in great numbers, and the latter have a very singular look, as the naked roots, several feet above the ground, give the trunk the appearance of standing on legs. On our return, we passed a hut, in which we saw an old woman certainly above a hundred years old; she looked as dry and shrivelled as a mummy: the weight of years had bent her, but by no means palsied her tongue; her talkativeness was indescribable; at the same time her toothless mouth poured forth a number of witty conceits, at which my companions laughed heartily. We saw many children; and this made the scanty population still more inexplicable, and, like the young plantations of cocoa-trees, indicated an entirely new settlement of inhabitants in these islands. One of my companions, an elderly man, who seemed to possess much natural understanding, pleased me very much by his behaviour. Lagediack was the name of my friend and teacher, for I really learnt more words from him in a few hours, than from others in several days. I had gained his confidence by some presents, and tried to learn from him many things respecting these islands, as he always knew how to make himself understood; he told me, for example, that this island is called *Otdia*, and the whole group is named after it. I found it easier every day to understand the language, because, as I soon observed, it is entirely destitute of connective particles. I invited my friend to visit me the next day, on board, using the following words: *Ildiu*, *Lagediack*, *Waedack*, *Oa* (morrow Lagediack come ship): he understood me perfectly well, answered *inga* (yes), and embraced me out of joy that I understood his language; but I believe that my joy exceeded his, particularly when I observed that I gained the confidence of the savages by my docility. I resolved to remain several weeks at Otdia, partly to explore by boats from hence the southern group of islands, and partly to make myself acquainted more closely with the language and customs of this remarkable people; for I think it very interesting, on discovering a country or an island, to study likewise the inhabitants, their manners and customs; besides this, I have had no cause, in the sequel, to regret the loss of time, as it was the very means of enabling me to make new discoveries.

On the 21st, I sent for water, which collects in Otdia in several pits, and is very good. In the afternoon we were visited by two boats, in one of which, was Rarick and his suite, and in the other the chief of Egmedio, a small island south of Otdia, which is distinguished by a grove of old cocoa-trees, situate in the middle of the island, and rising much higher than all the other trees. This, the high Bird island, and another south of our anchoring-place, are three fixed points which present themselves to the navigator when he reaches the southern point of the group. The numerous old cocoa-trees on Egmedio, make it still more inexplicable to me why they have but just begun to plant them in the other islands, when it might have been done so long before.

Rarick introduced to me the chief of the island Egmedio, whose name was Langin; he was above thirty-six years of age, of a middle stature, and very delicate; his whole body was tatooed, his dress tasteful, his behaviour modest, but I found him unreasonably timid. My friend Lagediack, according to his promise, came on board with Rarick, who had the courage to go to-day to the pig-sty, to examine the animals more closely, but was ready for flight at the slightest grunting. Langin, the most fearful of all, would not venture so near, but climbed along the ropes up the mast, and so looked down upon them. They were already so familiar with my little dog, Valet, that they began to play with him; but if, in his play, he began to bark, their friendship was at an end, and all my guests were in an instant in the shrouds, and they could not accustom themselves to his liveliness during my stay there. They were more pleased with another dog on account of his phlegmatic temper, which I had purchased in Beering's Straits; it was of the kind used in Kamtschatka, for drawing sledges; his coat resembled that of an ice-bear; being born in a cold country, he could not support the heat, and died of convulsions. After the eyes of the savages were in some degree satisfied with objects of luxury, the iron attracted their attention; a piece, for example, as large as an anchor or a cannon, seemed to them a prodigious treasure, and, with the constant cry of *Möll! Möll!* they examined every thing with the greatest attention. I gave them all presents to their entire satisfaction, but particularly, besides the two chiefs, to Lagediack, to insinuate myself more and more into his friendship. He was obliged to sit next to me, and I called up all my knowledge of the language to ask him, whether, besides this group of islands, he was acquainted with any others. My speaking and pantomime were long in vain: at length, however, he understood me, pointed with his hand to the south, saying, *inga eni cef cef,* (yes, islands there); and my joy was the greater, as I owed the discovery of an unknown group to my knowledge of the language. I now ordered the azimuth compass to be set up; all of them immediately crowded round the instrument, which they regarded with much attention; Lagediack, in particular, could not take his eyes from the magnetic needle, which turned without being touched, and he repeatedly asked me, how that could be? But how could I give him an explanation of the compass, even if he had been perfectly master of my language? how give him light in a thing respecting which I myself desired to be informed? That the box was to be turned while the needle kept the same position, he easily comprehended, as he directly discovered that the needle turned to N. and S. I begged him again to show me the situation of the unknown group of islands; he immediately took the compass, and, turning it upon the foot till the sights were directed to the part required, made me understand that the islands lay in that direction. The direction of the compass was S.W., which I immediately marked down on the tables. The writing was a new discovery, which excited their attention as well as reflection. I tried to make Lagediack understand that all we spoke was written down on the tables; wrote his name down, and said, that is Lagediack. He was greatly frightened to see himself represented by such singular figures, and seemed to fear that he would be obliged, by magic, to assume such a shape; the others laughed heartily at the comical Lagediack on the tables, while he himself stood in great uneasiness, expecting the terrible metamorphosis. I soon relieved him from his painful situation by effacing his name; he embraced me full of gratitude, and begged me to transfer Langin to the table; but the latter, who had looked at my conjurations timidly at a distance, on hearing this proposal, ran, with loud cries, to the other side of the ship, where he concealed himself; his comrades laughed at him, and my conjurations were finished for to-day. I tried to explain to Lagediack, that I wanted him to draw the whole island group of Otdia on the table. He took a pencil, and drew the group in a circle, under the lee of which he marked five channels, which he called Tier; he then turned the sights of the compass to the high island, in S.W., saying, *ef, ef, ruo Tier* (there two passages). This news rejoiced me the more as we should not be obliged to go the same way back; and, perhaps, nearer to us we might meet with a passage more convenient and secure than Rurick Straits. I then made Lagediack draw the other group of islands, which he called Enegup; and he again made a circle of seventeen islands, with several channels to the leeward, then pointed to the E., and gave me to understand, that if we left here at the rising of the sun, we could be there already at its setting. According to this description it must lie at a small distance from Otdia, and I did not doubt but that we should discover it without difficulty. My friend further told me that pandanus (*Bob*), bread-fruit (*Mia*), and cocoa (*Ni*), grew at Enegup; of the population he gave me but a poor idea, assuring me that there were only an old man and three women. The old man was therefore only a chief over his wives, or perhaps himself subject to a threefold dominion. When our guests left the ship, I made Lagediack a present of an axe, which Rarick almost envied him, and we parted on more friendly terms than we had hitherto done. Some of our gentlemen had been obliged to exchange names; Langin, who had conceived a particular friendship for Lieutenant Schischmareff, called himself after his name, Timaro; and Chamisso's friend was called Tamiso. They were not able to pronounce the names in any other manner.

I intended to wait for a day, when the wind should blow fresh, to explore the strait to the S.W., and as I had long had the plan of making a garden at Otdia, where I could sow the seeds from the Sandwich islands in the presence of the inhabitants, I went on shore at noon, accompanied by Mr. Chamisso, to select a piece of ground for this purpose. Near Rarick's habitation, we found an open place, in the vicinity of a reservoir of water, which seemed to answer the purpose; the earth was very fine, and I intended to work at it the next day. On our return, we saw another grave, resembling exactly that on the island of Ormed, planted all round with cocoa-trees, and we were informed that a Tamon was buried there.

The 22d, in the morning, we were visited by several canoes, which presented us with cocoa-nuts; on every visit, I tried to lessen their fear for the hogs, as I intended soon to bring them on shore. Immediately after dinner I went to the island with Mr. Chamisso, and several sailors, with shovels, to set about the work, that we might be able to finish the new garden before our departure. Rarick, Lagediack, and many inhabitants regarded us with looks of curiosity; the shovels excited their astonishment, but I strove in vain to make them comprehend my intention. As soon, however, as our work was so far advanced that we produced the seeds, light seemed to break in upon them. While some sailors were employed in rubbing the earth small, and digging it, others made a fence, at which Lagediack himself assisted, without being asked. By degrees, all the inhabitants of Otdia assembled round us, and gazed chiefly at the fence, with the use of which, these happy islanders were not acquainted. After a part of the garden had been worked on, and some seeds sown, we gave Lagediack to understand that they might expect from it plants and eatable fruits; and his leaping with joy convinced us that he comprehended us. He then declared aloud, our intention among the people; they all listened to him with the utmost attention, and when he had concluded his long-winded explanation, the joy was general. For to-day we concluded our work after a few hours, not to make it tedious to the spectators, and explained to Lagediack, that the garden belonged only to him and to Rarick, and that the fence was intended to prevent others from coming that way. He made this likewise immediately known; felt himself highly honoured by this distinction, and tied two different knots * of pandanus leaves, which signified his and Rarick's marks; they were hung to the fence, to show that both were possessors of the garden. During our work we had our tea-things brought on shore, and then went to Rarick's hut, where the kettle was already on the fire, round which the inhabitants assembled, looking at the boiling water, which they considered to be alive. Under the shade of a palm-tree the napkin was spread on the ground, and they all joined in the loud *O—h!* at this new wonder; but when we began to make the tea, there was no end of their talking and laughing, and they attentively observed all our motions.

* With these knots of pandanus leaves they distinguish their property; the owner is recognized by its make: we found them most frequently hanging on the trees.

The tea was ready, and their curiosity was without bounds, when they saw us drink it. I offered Rarick a very sweet cup, which he did not venture to take to his lips till after a great deal of persuasion. Unfortunately, the tea was very hot; he burned his mouth, and I but just saved my cup, which he was going to throw from him. The fright spread like an electric shock, and they were all prepared to run off. Rarick, at length, determined to taste it; the others looked at him with much astonishment, and when he found the tea agreeable, they all wanted to have some, and expressed, by a loud smacking, that they liked it; they were likewise fond of eating biscuits with it, but the sugar carried off the prize. All the islanders are great lovers of sweet things; and their chief food, which they draw from the sweet pandanus fruit, is probably the reason, that even children of ten years old have not good teeth, and that they have generally lost them all in the prime of life. This was the first time that the islanders had resolved to taste our food; a proof how much their confidence had increased; but, unhappily, their propensity to stealing showed itself at the same time, and Rarick himself was the first to give this bad example. The bright silver spoons caught his eye so much, that he attempted to conceal one in his girdle; but, when we observed it, he was prevented by a joke. A copper measure, which the sailors used in drinking water, was missed, and was found, after a long search, carefully hid in the bushes. As we had not hitherto lost any thing, I was convinced that this vice was foreign to the islanders; the more disagreeable was my surprise on this occasion. I expressed my dissatisfaction to those present, and ordered my people to be more careful for the future, that our friends might not be led into temptation.

On the 23d, in the forenoon, we were visited by Rarick and Langin, with a numerous train; they brought us cocoa-nuts, and were received with as much friendship as usual. They walked about quite unconstrained, but the cannons still attracted their attention, and they were of opinion that they served us in the place of their shell-horns, without thinking what a dreadful engine of death they had before them. One of Langin's attendants, having stolen a knife in the cabin, was obliged to return it: he was very much ashamed; Langin was very angry, and left us to go to his island, to which he invited us. In the afternoon we went on shore to finish the garden, and were struck at first sight by the devastation which the rats had made: even our presence did not seem to disturb them; some were rooting out the seeds, and others run away with their booty. I made Lagediack comprehend that the garden would be entirely destroyed if they did not place sentinels; and the thieves were soon driven out with stones and sticks. We repaired the damage that had been done, and planted the remaining part with water-melons, melons, maize, beans, peas, lemons, and yams. Lagediack soon perceived that all was good for eating, and the manner of preparing it had still to be explained to him. Having a baked yam-root still by me, I distributed it; and they found its taste so pleasant, that it very much increased their interest in the garden. We likewise distributed a number of seeds, which they might sow at pleasure. Mr. Von Chamisso was indefatigable here, as well as on all the other islands at which we touched, in sowing various kinds of seeds, and I am convinced that his labour will not have been fruitless. After having concluded our work, we returned to the ship with the consciousness of having done a useful act, and our trifling labour was rewarded by the gratitude of the islanders, who became more and more attached to us. The food of the islanders of this group consists, at this season of the year, solely of pandanus fruit; and they consider cocoa-nuts, which are here very scarce, as dainties. But as the pandanus contains very little nourishment, and is by no means in abundance, an idea of their frugal fare may easily be formed, which seems to agree with them, for they enjoy extraordinary health, and attain to an advanced and cheerful old age. A threefold increase of the population must produce a famine, which we may hope that our seeds will be the means of preventing. It is remarkable, that they almost totally neglect fishing; only a very few times, during our stay there, we saw the people on the reefs, angling for a species of small fish. A cock and a hen, the only remains of our poultry, I made a present of to Lagediack, whose joy compensated us for the loss of a dish.

The 24th of January. To-day, the forge was put up on land, as several iron articles required repair. The novelty of the scene attracted all the inhabitants, who beheld with astonishment the setting up of the machine; but, when the bellows began to work, the coals glowed, and the hammered iron threw out fiery sparks, the men seized their wives, the women their children, and all fled. Lagediack was the first to be convinced of his groundless fear; and, to make him acquainted with the use of the forge, a handsome harpoon was made on the spot, of which I made him a present, and caused him the most indescribable pleasure. Holding the harpoon over his head, he called with a loud voice to his comrades, who, encouraged by his example, came again together. I had another harpoon made in their presence for Rarick, and some fishing-hooks for my favourites, and their attachment increased in proportion, as every new specimen of our skill raised us in their eyes. As some days would elapse before the labours of the forge could be concluded, I left it on shore, under the care of the smith; and Lagediack promised to guard against any thing being stolen in the night. Mr. Chamisso remained also that night in Rarick's hut, to make himself acquainted with some of the customs of the islanders.

The 25th. The night had passed on shore without any one having ventured to approach the forge. In the morning, when the work had begun again, an old man suddenly stepped forward, quickly seized a piece of iron, and was going to run away with it; but his companions, who perceived it, pursued him, crying, *Cabuderi,* (steal,) and fetched him back, and were obliged to take his booty from him by force, as he would not resign it willingly. Without the least embarrassment, he resumed his place among them, was in a passion with every body, and immediately attempted to seize another piece, upon which it was found necessary to send him away. This old man, who came from another island, could not properly be called a thief, as he committed his depredation openly; for he evidently merely attempted to exercise the right of the strongest.

For some days past we had had a continual N. W. wind, with frequent torrents of rain, which prevented me from examining the channel pointed out by Lagediack.

The 26th. To-day, the hogs, to the sight of which, the islanders had become pretty familiar, were sent on shore, as a present to Rarick, near whose dwelling, a small place was fenced in. A sailor was obliged to remain on shore for some days, to instruct them in the management of the animals. From the sow, a speedy increase was to be hoped. Pleased as Rarick was with the present, he would not venture near them when he heard their dreadful grunting on landing, and the women, who had never been on board, and were acquainted with them only from the accounts of the men, ran away into the woods at the first sight of them. I made an excursion through the island with my fowling-piece, in the hope of shooting a land-bird, but did not find a single one, except a very few wild pigeons. Rarick and Lagediack accompanied me, without suspecting my motive; and, to give them a specimen of it, I showed them a sandpiper, about fifty paces distant from us, on the shore, and shot it; but I immediately had reason to repent of my want of thought, for both of them lay stretched out at my feet, lamenting aloud, and concealing their faces in the grass. After many assurances that they should receive no harm, they rose, trembling very much, and looking fearfully after my piece, which I had put up against a tree. The sight of the bleeding bird was not adapted to make a joke of it; they remained fearful and suspicious, and ran away, in a moment, when they thought themselves unperceived. It cost me much pains to regain their confidence, and I could never venture to show myself with my gun.

On the 28th of January, at seven o'clock in the morning, I left the ship, accompanied by all our scientific gentlemen, in two boats, provided with provisions for three days. We first sailed to Langin's Island, which we reached in about an hour, and were received by him in the most friendly manner. He immediately conducted us to his hut; his wife was obliged to prepare all kinds of things to treat us, and he did not cease expressing his pleasure at our visit. Langin, who showed himself a hospitable and obliging man, surpassed Rarick, who did not answer the first favourable impression; the chief feature in his character, covetousness, being disagreeably prominent. The population consisted only of Langin, his wife, and two men, who seemed to be subject to him. We already knew, from experience, that the whole group had a very scanty population: the southern part of it is entirely uninhabited. No other reason can be assigned for this scanty population, than that either lately a few people had been cast here from distant islands, or that they had, perhaps, voluntarily settled here, from over-populous islands. Langin conducted us about his possession, which is distinguished from the other islands, by the high cocoa-trees already mentioned. When we took our breakfast, in order to reach the place of our destination without delay, Langin was very much astonished to see us use knives and forks, and plates; he perceived that the flies disturbed us in eating, and immediately ordered one of his men to drive them away with a palm branch, an attention which agreeably surprised me in a savage.

At one o'clock in the afternoon, we reached the channel, next the high island; the breadth at the narrowest part, was a hundred fathoms; its depth was irregular, and decreased from twenty to five, and, in many places, to only three fathoms; the island consisted of pointed coral of various kinds. It was just at the time of ebb; the current ran with considerable violence out of the group: we suffered ourselves to be driven through the passage, into the ocean, and were scarcely through it, when we could no longer reach the bottom with the lead. To come through here with the Rurick, was indeed possible, but very dangerous. A violent wind, which just then arose, did not permit us to examine the second channel, which according to Lagediack's description, must lie in the west. I deferred it till a more favourable day, called this one Lagediack, and set out on our return, which was rendered very difficult by the contrary wind. It was impossible to reach the Rurick this evening, and we saw ourselves obliged to pass the night on the large island, south of Egmedio. Fortunately, we were here met by one of my favourites, the always friendly Labugar; he brought us cocoa-nuts and pandanus, and informed us, that we had landed on his island, which was inhabited only by his family, and an old man. We took up our quarters on the shore, to prepare our supper; and Labugar, with his old friend, entertained us till sunset. When we opened our eyes in the morning, Labugar sat with his family at our feet, patiently waiting till we should awake, to make us a present of cleaned cocoa-nuts. This tender attention moved and pleased me. In a walk, I found a piece of wood, clearly a part of a ship, in which there were still some old rusty nails. We reached the Rurick towards noon.

The 30th of January. — To-day, I sent a part of the crew on shore to fell wood, of which we should want a large stock, as we should not be able to procure any, either in Oonalashka, or Beering's Straits. When I came on shore, they told me, that a pail with iron hoops had been stolen. To prevent a second attempt of the kind, I resolved this time to be severe; I therefore seriously desired Rarick immediately to produce the thief and his booty; and he, astonished at my countenance, which he had never seen so angry before, assured me, that he had already heard of the circumstance, and had sent for the thief, who had fled to another island. I was satisfied with this explanation, but have since had cause to believe, that he was concerned in it.

The 31st. — I was much displeased to-day, on hearing that the pail had not yet been delivered. Rarick, who was employed with his men in building a boat, seemed confused on my inquiring for the thief; he turned angrily to one of his people, who, at the end of a very long conversation, sprung up, and ran into the bushes. "This is the thief," said Rarick; "he will immediately fetch the pail." I observed with great joy the expression of disapprobation on the faces of all present, and particularly Lagediack. In ten minutes, the thief returned with his prize, and though his obstinate physiognomy displeased me, he was excused from punishment. After declaring, that every future theft should be severely punished, I returned to the Rurick, where we were just sitting down to table, when Labugar and Rarick, with a very talkative old woman, who belonged to his family, came on board. She, the most unamiable of her sex, was the first woman of these islands who had ventured on board. We invited our guests to dinner. The two men readily went into the cabin, but the old woman took her seat on the quarter-deck by the window, through which she looked down upon us. The merry Labugar ate every thing that was given him, and at every different dish, asked, What is that? but had swallowed it before he could be answered, and took care to promote digestion by laughing heartily; even the salt meat, which we told him was of such animals as the two they had on land, pleased his palate. Rarick was at first moderate; but when he saw how much Labugar relished it, he took care to eat heartily. The old woman carefully preserved every thing that was handed to her through the window, opening her mouth wide, to shew us that she had no teeth ; if this want prevented her chewing, her mouth was the more active in talking. Labugar relished wine extremely ; he was pleased to feel how it came down into his stomach, which he held, that it might not run out : after having drank two glasses, he played the most foolish tricks. To treat our guests very highly, we had a concert after dinner. Three sailors, all novices in their art, came into the cabin with a violin, flute, and tambourine ; it seemed, however, to our savages, as if the immortals had descended to afford poor mortals a pleasure. The violin excited the greatest astonishment ; they would not believe, that the sound was produced by the bow, and took care not to come too near to the magic instrument. After they had received some presents, they left the ship in high spirits. In the afternoon, we rowed on shore, and I saw, to my great joy, that some of the seeds were already springing up in the garden. Chamisso pointed out to me the small gardens which the inhabitants had made in several places. We could say, in the literal sense of the word, that the seed had fallen in good ground, and promised fine fruits for the future.

The 2d of February.— After we had had a strong east wind with rain, it cleared up towards evening, and I took advantage of the fine weather to visit my friend Lagediack, from whom I wished to obtain the geographical situation of the group of islands, called Eregup. He really invented a very clever method of giving me a correct idea of it: he drew on the sand a circle, nearly in the form of the group Otdia, placed round the edge of it large and small stones, which represented the islands; and after having marked the channels, he said, "That is Otdia." The island-group of Eregup, which he explained to me in the same sensible manner, I might reach, he thought, in a day's voyage to the south-west. I have in the sequel found his information to be perfectly correct. I then endeavoured to learn from him, whether we should meet with islands on sailing N., E., S., or W. To my great joy, he understood me, fetched a number of stones, and began to mark north of Otdia three somewhat smaller groups; the intervals were all one day's voyage; but the last, two: he called these Ailu, Udirick, and Bigar. At the distance of a day's voyage, he marked another to the north-west, which he called Ligieb. When he had described those to the north, he went to the south, and marked there five groups; the spaces between which were likewise from one to two days' voyage, and called them Kawen, A-ur, Mediuro, Arno, and Mille. To learn more of these groups was beyond the reach of my scanty knowledge of the language; but this information was very welcome to me, and procured Lagediack many presents: I therefore determined to leave Otdia as soon as possible, to pursue our discoveries. According to our calculation, Kutusof's and Suwarof's islands lay in nearly the same longitude as Otdia; the latitude was only $1\frac{1}{2}°$ different, and I made no doubt that they were among the groups marked in the north.

On the 3d, at six o'clock in the morning, I dispatched Lieutenant Schischmareff in the long-boat to examine the second passage: he came back in the evening, with the news that the channel was very safe, and at the narrowest part 150 fathoms broad; he had found the middle to be unfathomable, and near the reef 11 fathoms deep. Highly pleased at this information, I ordered the Rurick to be ready to sail in a few days from Otdia. A very disagreeable circumstance occurred to-day at dinner: Lagediack was our guest at table; his companion (he who stole the pail) was refused admittance into the cabin, and he was obliged to sit on the quarter-deck at the window, and content himself with looking on. Full of compassion, Lagediack reached him several pieces; but the bright knives pleased him more than all the food; he begged to have one to look at, and hid it in his girdle. As we purposely pretended not to observe it, in the hope that he would return it, I remained silent; but when Lagediack was going on shore, and the thief about to leap into the canoe, I gave the signal agreed upon; four sailors seized him, and after having taken away the knife, they laid him down and severely punished him. Lagediack, greatly astonished, entreated for his friend, repeatedly exclaiming, *Cabuderi emo aidaro*, (steal not good;) the latter, after his punishment, got into the canoe, and seemed to regret having been deprived of the knife. This circumstance was much laughed at on shore, and Rarick and Lagediack visited me again in the afternoon, bringing presents of cocoa-nuts and roasted fish, as a proof that they did not disapprove of what I had done. I now told my friends that we should soon leave them, which seemed to cause a disagreeable surprise. Lagediack insisted on knowing where we were going, and whether we would soon return. My answer, that we were first going to sail to Eregup and Kawen, to give the people iron, and then entirely leave their islands, affected them all very much. Lagediack particularly testified his affection by repeatedly embracing me; he infected me with his feeling, and I felt as if I were parting from an old friend when he left us with Rarick after sunset.

The 6th of February.—The news of our departure, which was speedily circulated, brought us a number of farewell visits. To-day, being the last of our stay, my best friends did not quit the ship, and I delighted Rarick and Langin with a piece of sail-cloth for their new boats. In the afternoon I was visited by the venerable old chief of the island of Ormed; we were all very partial to this old man, and he received many presents; an old coat, with bright buttons, which we gave him, he put on the instant he received it. For the last time, I to-day enjoyed, on shore, the sight of our garden, where every thing throve admirably; and Rarick and Lagediack left us after sunset, with tears in their eyes.

OBSERVATIONS MADE ON THE ISLAND OF OTDIA.

The mean of our daily observations gave, for the latitude of our anchoring place, - -	9° 28′ 9″ N.
The mean of 300 distances of the sun and moon, taken on different days, gave for its longitude - -	189 43 45 W.
Variation of the magnetic needle	11 38½ E.
The mean height of the thermometer - - - -	82° Fahrenheit.
The mean height of the barometer, which during our whole stay varied but a few lines -	29, 7 inches.

As the low land here has no influence on the atmosphere, the barometer falls and rises as uniformly as it generally does between the tropics. The mean of our observations at Otdia gave for the time of high water, at the new and full moon, two hours thirty minutes: the greatest difference in the height of the water was seven feet.

To this group, which consists of sixty-five islands, I gave the name of Romanzoff.

CHAP. XII.

FROM RADACK TO THE ST. LAWRENCE ISLANDS.

THE 7th, at day-break, we weighed anchor to leave the beloved Otdia, where we had passed so many happy days among the uncorrupted children of nature. The sails were spread, the weather serene, the wind favourable, and the friendly islanders, who were assembled on the shore, cried to us their last adieu. I now took my course to Schischmareff Strait: we were not endangered by any banks, of which we only saw two at some distance. A canoe followed us from the island of Ormed. At eight o'clock we reached Schischmareff Strait, which is in every respect preferable to Rurick's Strait, as it is much broader, and permits the mariner to sail in and out, with the usual monsoon, without even being obliged to tack; Rurick Strait, likewise, is not so easily found, as nothing is seen there but a continued reef, which is always alike: here, on the contrary, the passage between the two islands is to be seen, even at a distance. In the middle of the strait, the sailor at the mast-head cried out, Land! In S. by W. we saw two small islands; the group of Eregup was there also visible.

I immediately took some altitudes of the sun,

and found the longitude, according to the chronometers, 189° 50′. The latitude, according to the ship's reckoning, deduced from our anchorage, was 9° 24′ 57″ N.

We had not at all expected to come so soon in sight of the group of islands; it proved that Lagediack's day's voyage was no standard for us; and we concluded that the other groups were nearer than he had fixed. As we saw Eregup before we lost sight of Otdia, it was easy so to unite by angles both the groups, that Eregup had its exact position assigned it on the chart.

At ten o'clock we had reached the northern point of the group of Eregup, which consisted only of coral reefs, and directed our course to the west, and sailed through the channel formed between Eregup and Otdia, to get under the lee of the former. The current in the channel caused a loud roaring; the waves towered like breakers over a shoal; I had the lead thrown, but did not reach the bottom at one hundred fathoms. At noon we had sailed round the northern point of Eregup; we were under the lee in calm water, and sailed along the western side, at the distance of a mile, our course always directed to S. E., where the group took this turning. After a good observation, we found ourselves to be in latitude 9° 9′ 6″ N., longitude, according to the chronometers, 190° 2′ 47″. The wind had now turned to the E., and we were obliged to tack to reach the southern point of the group.

We were soon enabled plainly to overlook it, and found it considerably smaller than Otdia. Its length is twenty-four and its breadth four miles. The whole circle consists of one reef, and contains but very few islands, which are clearly laid down in the chart.* At four o'clock we were near the southern point of the group, which closes with the largest island; and this, probably, bears the name of Eregup; at least it was the only one on which we saw cocoa-trees and inhabitants. I could not now avoid giving credit to the assertion of Lagediack, that it was inhabited by only three people, as even the appearance of our ship did not attract more to the shore. I dispatched Lieutenant Schischmareff to examine a passage near to us; but he soon returned with the news, that it might perhaps be navigable, but dangerous, on account of the numerous turnings, and could not be passed, except with a west wind. On this information, I gave up the further examination of this group, which seemed to me not worth spending much time upon: the survey was concluded, and at seven o'clock we sailed round the southern point. We now tried to get the weather gage to the N., to sail with a S. E. course direct to the island group of Kawen, which, according to Lagediack's statement, must lie to the E. We could not find the strait on the north point of Eregup, which he marked. I called this group after our former minister of the marine,

* Vide Chart of the chain of islands of Radack and Ralick.

Tschitschagof. At sunset we stood off from land, and tacked during the whole night under few sails, with fine weather, and a moderate wind from E.N.E.

The 8th of February. — The south-east part of the group Eregup lay before us at day-break in N.W.; we had to combat with a current from N. during the night; all the sails were spread; the wind allowed a northern course, and, at seven o'clock in the evening, we saw to the N. the high islands of the group of Otdia, near which is Lagediack Strait: at the distance of three miles to the left of us lay the group of Eregup. A good observation at noon gave us the latitude 9° 9′ 49″ N., longitude, according to the chronometers, 189° 51′ 14″. We found that the current had carried us, since yesterday evening, six miles and three quarters to the S. The whole day and night were spent in tacking.

On the 9th, we lost sight of Otdia; the weather was serene, and we constantly tacked. In the forenoon, the moon appeared, and we immediately took a number of distances between her and the sun, from which the longitude was calculated, and reduced from noon, and gave 189° 20′ 20″. The chronometers gave for noon 189° 26′ 43″; the latitude, according to observation, was 8° 53′ 16″. The current had carried us since yesterday at noon, nine and a half miles to S.E. 28°, and on this account we had not been able to reach the point from which I hoped to make the island group of Kawen by a S.E. course. The wind was violent, and we tacked during the night.

On the 10th, at six o'clock in the morning, we reached the point which we so much wished for; all the sails that the high wind permitted us to carry, were spread; we advanced rapidly, and hoped every moment to descry land, while the high irregular waves kept the Rurick in violent motion. At noon we were in latitude 8° 55′ 52″ N.; longitude, according to the chronometers, 189° 20′ 13″ W. The current had carried us, in twenty-four hours, twelve miles and a quarter to S.W. 88°. We had scarcely laid our instruments aside after the observation, when land was descried from the mast-head in the E., at a distance of ten miles. Some of our gentlemen had already given up the hope of finding the group, in the existence of which they had begun to doubt. Lagediack had laid down its situation pretty accurately; it was forty-five miles distant from Otdia. We quickly advanced towards the group, which, with its islands, covered with high palm-trees, afforded a more pleasing prospect than Eregup. When we were off the W. point, we saw in S. and S.E. a chain of islands which extended so far that it seemed to be lost in the horizon. On the western point is the largest island of the group, which we afterwards learnt was called Kawen. At four o'clock in the afternoon, we were under the lee of the group, which protected us from the waves of the high sea, and enabled us to follow up the chain at a distance of half a mile to S. After we had sailed along it nine miles, and the sun was near setting, I gave up further examination for to-day; the ship was put about, and we sailed back the same way by which we came. We had seen two channels between the reefs; the larger one near to the island of Kawen, and the second a little more south; and I determined, if possible, to penetrate into the group next day. On our voyage back, we saw a great number of people on Kawen, who were admiring our ship; we therefore hoped to meet here with a more numerous population.

On the 11th, at break of day, the wind became violent, but the weather remained serene; I therefore did not abandon my determination of penetrating the group. At six o'clock we saw the island of Kawen, and at half-past seven were in smooth water before the strait; which lies close to this island. The wind was now so violent that the top-sails ought to have been double-reefed; but as the direction of the passage made it impossible to enter, except by tacking, I did not dare to carry little sail, and gave orders to take in only the top-gallant sails. We had, however, in the mean time, approached the entrance within a fathom, and could clearly see whether there was enough depth for our ship. It must appear strange to the mariner, that we should attempt to measure the depth of the water by our eyes; however, as I have said before, the water between the coral groups was so transparent, that even when sailing we could see the bottom at eight fathoms, and besides, our eye was so practised in this respect, that we could know the depth of it by its colour. When we approached the strait, we did not find it so broad as it had appeared at a distance. Two shoals concealed under the water, between which we should be obliged to tack, formed a channel of less than a quarter of a mile, and though the undertaking was dangerous, with such a high wind, we determined boldly to brave the danger. While we were examining the channel, two boats, notwithstanding the high wind, each manned with seven islanders, put off from the eastern part of the group, and sailed towards us, but remained at a distance of two hundred fathoms, and tacked as we did. We were just admiring their skilful manœuvring, and the possibility of carrying such a large sail in this wind, when a sudden gust upset one of the boats. We saw all the people swimming about; one part of them, who appeared to be women, quickly got upon the keel, the others tied cords round their bodies, and swimming, towed the boat on shore. Meantime the other canoe, unconcerned at the misfortune of its comrade, sailed directly before the wind to E., where it landed. Soon after this two larger boats came up to us from the island of Kawen, and kept constantly under sail, the people making all kinds of movements with their hands, and calling out something to us, which we could neither understand nor hear for the roaring of the wind. There was not the least difference in their boats and costume from those of Otdia, and we could no longer doubt that it was the same nation. They likewise paid no attention to their fellow-islanders in distress, and did not offer to assist them, though they were still half a mile from shore, and had enough to do to reach it. We had now to exert our skill to carry the ship safely through the channel, as we were in danger of being wrecked at every unsuccessful turn; and though a heavy rain sometimes intercepted our view, we had reached, at nine o'clock, the interior of the group, where we had sufficient room to tack, as no coral banks were visible. We found the depth, over a bottom of living coral, in the middle of the channel, twenty-three fathoms, and it regularly decreased to about five fathoms on either side. But we had scarcely entered the basin of the group, which had a depth of from twenty to thirty fathoms, when we were compelled, as the wind increased in violence, to reef the top-sails; yet we gained much, notwithstanding the few sails, by tacking, as the water was as smooth as a mirror. At three o'clock in the afternoon, we reached the chain of islands lying N. of the strait; Lieutenant Schischmareff, going out on board a yawl, discovered an anchoring-place, and in half an hour we cast anchor, at two hundred fathoms from a small island, and in twenty-three fathoms water, on fine coral sand. The island of Kawen was distant from us five miles in S.W.; the boats of the inhabitants had left us on penetrating the group. Soon after we had anchored, I rowed to the small island, accompanied by some of our gentlemen; it was of the same nature as those of Romanzoff's group, and seemed to be uninhabited: there were only a few small empty huts, which probably served the islanders for their abode when they fished. We also found rats here in great numbers.

The 12th of February. — Though our anchoring-place, with considerable depth of water, and but little protection against the high waves, was not convenient, I resolved to remain here to-day on account of the fine weather, to determine this point astronomically. Towards noon, two large boats from the eastern part of the islands came up so close to us that we could count thirteen men and three women. The word *Aidara*, which we called to them as the best recommendation, had the desired effect; the savages were greatly astonished at it, unanimously repeated the salutation, and made preparations to approach the ship. We now felt the great use of our hardly acquired philological knowledge; for when we invited them to our ship, in their own language, they did not hesitate a moment, and only begged that we would throw them a rope. We immediately complied;

the rope fell into the water, and one of the islanders instantly leaped in, seized it, and swam to the boat, to which he fastened it. After this work was finished, during which many compliments passed between us, two savages came on board, one of whom I recognized to be the chief by his distinguished ornaments. His head was adorned with white feathers, besides an enormous wreath of flowers; his neck with several skilfully wrought bones, and his body wrapped up in finely-braided mats. He was a handsome man, tall and robust, and his whole figure was very prepossessing. As soon as he came on board, his first care was to be acquainted with the Tamon: and when I was introduced to him, he came up to me, presented me with a cocoa-nut, and placed his elegant wreath on my head, pronouncing the word *Aidara*, which he frequently repeated. Upon this he told me, that he also was a Tamon of the island of Torua, lying to the east, where I might visit him. That I might entirely conciliate his confidence, I offered to change names, a proposal which was favourably received; with expressions of the greatest joy it was announced, that their Tamon was called *Totabu*, and I *Labadeny*. No friendship can here be permanent without presents: I therefore gave a solid basis to this newly-concluded alliance, by various iron articles, which my name-sake joyfully received, but did not give them to his treasurer till he had sufficiently feasted his eyes with them.

I have forgot to mention, what I already observed in Otdia, that every Tamon was always accompanied by a treasurer, who took charge of all the presents, and seemed to be at the same time chief favourite. Labadeny's treasurer, a buffoon, was lean, quick in his motions, had a very animated look, spoke a great deal, accompanying every word with violent gesticulations; in short, he was the very reverse of his austere master. The friendly reception inspired our guests with so much courage, that they fearlessly went about on the deck to satisfy their curiosity; the treasurer, in particular, took the most lively interest in every thing, would know the use of every thing, and did not cease exclaiming at the quantity of iron, *Möll! Möll!* He was immediately as familiar with us as if he were among his most intimate friends; if he wished to have an explanation about any thing, and I was near, he ran up to me, seized me, and dragged me away; he communicated all his remarks with great vivacity to Labadeny; and, not satisfied with this, every time he had seen any thing remarkable he sprung on the gallery of the ship, related to his fearful comrades in the boats the wonderful things he had seen, while they listened with open mouths. A barrel of water on the quarter-deck did not remain unnoticed. He asked what it was, and when I told him that the water was for us all to drink, he made the strangest capers, and did not lose a moment to communicate this news to his companions. He certainly did not do this without exaggeration, which I concluded from the astonished faces of his auditors; particularly a lively old woman, who seemed to be of high rank, and quite enraptured with his accounts, as she probably had not for a long time heard so much news as at this moment. Before Labadeny parted I told him that every one should receive iron in exchange for cocoa-nuts; and we separated good friends, after he had again invited us to Torua. We were now quite convinced that the inhabitants of Kawen and Otdia were one and the same people; their manner of tattooing, of dressing, was perfectly alike, only that a greater degree of luxury evidently prevailed here. The very large rolls in their ear-holes were ornamented with tortoise-shell; the mats of their dress were all new, and their necks hung with an extraordinary profusion of ornaments: the white feathers also, which have a very good effect in their black hair, I did not observe in Otdia. Kawen is the largest island of the whole group; it is two miles and a quarter in length, and three-quarters of a mile in breadth. The wind blew violently in the afternoon from E.N.E., and increased to a storm during the night, with frequent rain. It is remarkable that the barometer stood higher in this group than in Otdia; there its greatest height was 30,00, and here it suddenly rose, even in bad weather, to 30,80.

The 13th of February. I could not execute

my design to-day of sailing farther to the east, as the wind still continued to blow violently, and the sky was at the same time so gloomy that we could scarcely see the horizon. It was not till the evening that it cleared up, and we hoped to be able to continue our voyage to-morrow.

We found the latitude of our anchoring-place to be, - -	8°	54′	21″	N.
Longitude, according to the chronometers, - -	189	7	59	W.
Variation of the magnetic needle,	11	30	0	E.
Longitude of the middle of the island of Kawen, - -	189	11	27	W.

The 14th, at six o'clock in the morning, we were already under sail, with a fair wind and serene weather. Labadeny, who was also unable to sail yesterday, on account of the wind, tacked in his boat in our company to the east. On sailing past the islands we saw them overgrown with palm-trees which are so rare in Otdia. Many people wandered along the shore; columns of smoke every where rose up into the air; boats rowed to and fro; and the whole scene appeared lively and animated, contrasted with the death-like silence which prevailed at Otdia. When we had reached, by tacking, the middle of the basin, we could clearly overlook the chain of islands which forms the southern part of the group. The depth was here thirty-two fathoms: the bottom consists of living coral, and it is only near the islands that a fine coral sand is found: no coral banks here impede the navigation as at Otdia. In the afternoon we reached the little island, called by the natives Tjan, beautifully planted with cocoa-trees, where numerous huts and people indicated great population. We approached it within a couple of hundred fathoms, and found a convenient anchorage covered by a reef towards the east, where we dropped anchor, for the purpose of stopping a day to become acquainted with the inhabitants. Labadeny, who had followed us in his boat, now came on board, and urged me to sail to the island of Torua, lying more to the east, because, though this one also belonged to him, it was not his usual residence. I pacified him by a promise to go there to-morrow for certain, and now proceeded on shore with him. The nature of the coast is such, that you cannot get to it without wetting your feet Labadeny had the civility to carry me upon his back; and the strange sight of their chief panting for breath under the weight of a white man, drew many islanders to the spot. After he had happily set me down on the shore, he made a long speech to his people, of which I could only understand that he announced me as a mighty *Tamon;* after this he took hold of me, and conducted me into the interior of the island, into a very spacious hut. All the inhabitants collected round us, among them some pretty girls, adorned with flowers; and I observed with pleasure, that they here fulfil their destination; for they were the general ornament of the young girls. Old and young now strove to present us with cocoa-nuts for refreshment; but the treasurer took it upon himself to entertain the company. He had a great deal to relate of all he had seen on board the *Ellip Oa*, (large boat,) and what an immense quantity of *Möll! Möll!* there was there. At length he produced the presents which Labadeny had received, to excite his auditors to greater astonishment. The friendly, sensible behaviour of the islanders, made their society very agreeable, particularly as, relying on their numerous population, they are more courageous than the inhabitants of Otdia. Their curiosity was as lively as there, and I willingly fulfilled all their wishes to examine every thing, except undressing myself to please their curiosity. We now made an excursion into the interior of the island, which is, at the utmost, a mile in length, and a quarter of a mile in breadth. Considering its small extent, it is very populous; to judge from the number of large huts, from fifteen to twenty families, and that is almost as much as the whole group of Otdia contains. Cultivation has so far arrived to a high degree, as only useful trees, such as cocoas, pandanus, and bread-fruit are suffered here; every possessor has surrounded his small grove with a cord from one tree to another, probably not to encroach on a stranger's property. This precaution, which would appear laughable in civilized countries, is here sufficient to protect every man's right. The whole island has the appearance of an English park, here and there winds a path over-shaded by the spreading branches of the large bread-fruit tree; the majestic palm stands close to the pandanus, which rests on its high roots as on feet, and every where the wanderer finds a retreat against the scorching rays of the sun. The huts lie scattered under the shade of the bread-fruit trees, and we did not pass one, without the hostess inviting us to sit down on neat mats, and partake of some refreshments. We looked about in vain, as we had done in Otdia, for an object from which we could form some idea respecting their religious belief, but we could discover no traces; probably they worship an invisible being; for a nation of such good moral conduct, cannot be supposed to be without religion.

Their whole riches, in poultry, consisted of two fowls of a small kind; Labadeny offered them both to me, as a valuable present, but I was far from robbing them of their little treasure, which I was, unfortunately, not able to increase, as I had no more fowls. Chamisso discovered three kinds of *taro*, which had been carefully planted in a damp situation; but the quantity was so small, that these roots cannot serve for ordinary use, but only be regarded as dainties.

When I was going on board, the inhabitants brought us so many cocoa-nuts, that the boat was half filled with them; for which I returned iron. Labadeny again asked me from whence I had come, and where I was going. My answer, that I came from Otdia, and was sailing to Aur, astonished them all very much, as they could not conceive how I should be so well acquainted with their islands.

The bad weather prevented me to-day from leaving Tjan, at which Labadeny, who brought me early in the morning some cocoa-nuts, and a few small baked fishes, was much rejoiced, and begged me to go on shore in his boat. The wind blew hard when we left the ship, and I soon learnt how easily such a canoe upsets, if it is not kept in due balance. One of the savages, in spreading the sail, had stepped too far on the loof: and we should certainly not have come off with only the fright, if the rest had not immediately thrown themselves to the other side. The friendly reception on land compensated for our disagreeable trip; we were conducted to a hut, which sheltered us from the bad weather, where we sat very domestically on neat mats. I met here the old woman, whom I had seen on Labadeny's first visit to the ship, and learnt that she was his mother; she had conceived a very great affection for me, called me only Labadeny, and talked to me continually, which I so far liked, as it accustomed my ear to the pronunciation. I observed, with pleasure, that these islanders are as clean in their persons as in their domestic arrangements; only, that they, like the Otdians, have a very nasty employment. A pretty young woman, who sat opposite to me, laid her husband's head in her lap, to cleanse it from the vermin, and when she found any thing, she immediately bit it, and then swallowed it; it appears, that the women have the exclusive privilege of this chace. The inhabitants of the Aleutian Islands are said to have formerly had this custom; and on the N.W. coast of America, races of people are still found, who continue to practise it.

A fire was lighted before the hut, to prepare the dinner, and I saw, on this occasion, for the first time, how they produce it. A piece of soft wood is rubbed with a piece of hard-pointed wood, till a kind of groove is produced in the first; the dust caused by it supplies the place of tinder; and when the rubbing has increased the heat so as to take fire, they blow it up, and hold dried grass over it. Before the cook prepared the meal, he washed his hands; after he had finished his work, he spread a mat, set before us boiled fish of a good taste, and baked bread-fruit on clean leaves. Labadeny first tasted every thing, made a short speech, of which I did not understand any part, and then gave me the sign to begin. Only he, his old mother, and myself, were allowed to sit at our table; the treasurer, and another man, with whose office I am not acquainted, had permission afterwards to feast on the tails of the fishes eaten by us.

When we went on board, the inhabitants presented us so liberally with cocoa-nuts, that every one of the crew received one daily for a long time.

Since we came into the group of Otdia, our stock of water from Woahoo, has not been diminished: we found here likewise very good water in pits. M. Chamisso has distributed seeds among the inhabitants to-day, and sowed some himself.

We found the latitude of our anchoring-place - -	8° 52′ 39″ N.
Longitude, according to the chronometers - - -	188 58 29 W.

High water sets in, in the new moon, at thirty-five minutes after four, P.M., and rises to five feet.

The 16th of February. The weather was serene, we spread the sails, keeping close to the chain of islands by a south course. We were delighted at the admiring crowds, who, attracted by the sight of our ship, ran together on all the islands. After we had gone seven miles, a large, well-manned boat put off from one of the islands, which we afterwards learnt was called Olot; they sailed up to us, showed us cocoa-nuts, and when we pursued our course without attending to their signs, they followed us: Labadeny's boat was also visible at a distance. We had now before us a pretty large island, from which the chain took its direction to the S., and I resolved, as the group appeared to end here, to anchor, because I thought it important to determine its situation astronomically. By the bend which the island forms, we lay near it, protected from the wind, and had twelve fathoms' depth. Fortune had just directed us to Labadeny's residence, Torua. The large boat which had followed us from Olot came alongside of us, and the islanders came on board without any hesitation on hearing themselves addressed in their own language. The leader of the canoe, a young man about twenty years old, having first asked for the *Tamon*, presented me timidly, and in a friendly manner, with some fruit, and I soon gained his confidence by some small presents. I learnt that his name was Langedju, and that he was chief of the island of Olot; he was more tattooed, and therefore more distinguished than Labadeny; he had likewise a greater profusion of ornaments than he had; his face was animated and full of expression, and his whole behaviour pleased me much. The ship, with the quantity of iron it contained, was again an object of admiration; one of the savages even attempted to possess himself of some of the *möll*, but he was discovered, and the chief with much warmth gave orders not to touch any thing. Langedju proposed to me an exchange of names, which I was obliged to accept, not to affront him, though Labadeny would take it very much amiss; in fact, when he soon after arrived, he instantly perceived the change of names, and could not suppress his rage. Langedju was cunning enough to pretend to take no notice; I tried to conciliate my old friend by presents, but he would have nothing whatever to do with his rival. In the afternoon I went on shore with both chiefs, where we were kindly received by the inhabitants. Torua is twice as large as Tjan, but, in proportion, not so populous, though the island appeared to be very fertile. Immediately on landing, Labadeny left me; Langedju, however, accompanied me, with two of his subjects, in a promenade through the island; he was in a very good humour, laughed a great deal, and made game of Labadeny. After a long walk, in which I did not meet with any thing remarkable, except that some fowls ran about wild, I set myself down on the shore, and embraced the opportunity of extending my knowledge of the geography of this chain of islands. I drew on the sand the groups with which I was made acquainted by Lagediack. Langedju was greatly astonished that I was so well acquainted with their names, but found their situation not quite correct, and drew it himself. He placed himself towards the N., began with the group Bigar, and continued to the S., till the map was finished; in the number of groups it coincided with Lagediack's, but not quite in the direction. He had marked, besides the chain, two single islands, of which I took the one that lay E. of Ai-lu for New Year's Island; the other lay at the distance of a day's voyage to W.; the eastern one he called Miadi, that to the W. Temo. The map, as was afterwards proved, was very correct; for, as I discovered all these groups, I have accurately copied it in my note-book. On my return, I fell in with a company who were sitting round a fire, and boiling something in cocoa-shells. I accepted their invitation to sit down, and I now saw that they were making into powder rotten cocoa-wood, and boiling it in water to a thick pap; of this they made little cakes baked in leaves, but which I found to be quite tasteless. I was more pleased with the tender attention of the parents towards their children, which I remarked here likewise in a very high degree. I remained on this island till the 19th of February; Langedju scarcely ever left the Rurick, and brought me once some taro-root, upon which he set a very high value. We daily received a number of visitors, but Labadeny did not make his appearance. We changed iron for cocoa-nuts; the sailor who had this office was held in great esteem by the inhabitants; they embraced and kissed him continually, probably in the hope that their fondness might be taken into the account.

The mean of our observations gave for the latitude of our anchoring-place - - -	8°	43′	10″	N.
Longitude, according to the chronometers - - -	188	50	25	
Variation of the magnetic needle	10	50	0	

Our distance from the island of Kawen was, in a direct line, twenty-four miles.

The 19th of February, at six o'clock in the morning, we were under sail, and ran along the chain, which consists entirely of small islands, to the south. After we had sailed ten miles, it suddenly turned to S. E.; in this direction we discovered a pretty large island, and perceived that we were on the S. E. point of the group Kawen, which here first turns to W. and then to N. W., and thus forms a bay. I directed my course to the largest island, which we afterwards learnt was called Airick; and as we were approaching it, more islands were descried from the mast-head, over the reef to the south, which we soon recognized to be the group of Aur. At nine o'clock we cast anchor sixty fathoms from the island of Airick, in eight fathoms' water, and found this anchoring-place to be in every respect very excellent. Airick is about the size of Torua, but afforded a more beautiful prospect than any of the islands which we had yet visited. The whole shore was thickly planted with palm-trees, under whose shade many habitations were scattered; some boats were under sail, others lying at anchor, as if we were in a much frequented harbour. As we were so near the shore we could observe all the movements of the savages, who assembled in crowds to gaze at the wonderful great *Oa*. While we were putting the ship in order, some of our gentlemen went on shore, and we saw them received with palm

View of the Island of Airick, belonging to the Group of Kawen, in the Radack Chain.

London, Published by Longman, Hurst, Rees, Orme & Brown, 1821.

branches and cocoa-nuts. M. Chamisso soon returned, delighted at his kind reception; he brought with him a youth of eighteen years of age, who was announced as the Tamon, and whom the people followed in several boats, when they saw their young chief rowing from shore. We now had a numerous company, and were presented with cocoa-nuts in abundance, for which old pieces of iron were joyfully received. The Tamon, who was immediately introduced to me, took a lively interest in every object that surrounded him: an old man, who appeared to be his Mentor, did not leave him a moment, and both at last hit upon the idea of measuring the length and breadth of the ship, as well as the heighth of the mast, with a cord, which was carefully preserved. Seeing two gentlemen fencing on the quarter-deck, he begged us to give him a foil, and showed much aptness in the lesson. In the afternoon the Tamon seated himself next me in the boat, and we rowed, accompanied by all the canoes, on shore, where the people assembled to look at the Tamon of the ship; but my young friend, who did not quit my side, immediately conducted me to an elderly woman, whom he introduced to me as his mother, and queen of the island; she sat on a mat before a pretty house, surrounded by three very ugly, old and stately dames. I was obliged to sit next to her, and the people formed a close circle round us. Duly appreciating the place of honour, I exerted all my eloquence, but it was in vain; I obtained no answer; and, diligently as her eyes wandered over me, the royal mother did not deign to speak; and I was at length convinced that her high dignity imposed silence on her, particularly as the ladies of honour, in direct contrast to their mistress, chattered without ceasing. A present which I made the queen was received with a gracious inclination of the head, but she did not touch it; the ladies of honour took charge of it, and after the presents in return, consisting of some cocoa-nuts and two rolls of mogan had been laid at my feet in profound silence, the queen retired to her house, and the audience was at an end. The young Tamon then conducted me to a tolerably large house, supported on four posts, where I found an assemblage of young and profusely ornamented ladies; one of them, the sister of my companion, sat apart; I was invited to sit next her, and the people again formed a circle round us. In this visit to the princess, there was not so much ceremony as with her august mother; she asserted her right to speak, and was much pleased when I said any thing to her in her own langage; the people too were permitted to take a share in the merriment, and to joke. The princess ordered a pantomime, with songs, which is called by the inhabitants *Eb*, to amuse me. Two of her play-mates sat themselves by her, the one beat a drum, and the other joined now and then the solo song of the princess, which resembled screaming. The name

Totabu was frequently repeated, and I only regret that I could not understand the words. The pantomime would perhaps not have been much amiss if they had not, in the heat of action, distorted the eyes and twisted the neck, and at the same time gesticulated so furiously, that the foam stood in their mouths. On departing, I presented the princess with a silk handkerchief, and some other trifles; at which she was so delighted that she did me the honour to present me with her own shell wreath. The distinguished chief and his sister were not yet tattooed; probably the operation is not performed so early, as it is concluded in this island in one operation, and not, as in the Marquesas, in several. Langedju told me, that after tattooing, the body swelled very much, and suffered a great deal of pain: and really this pain must be very considerable; the mate of Captain Krusenstern, a tall, robust man, fainted when he had his arm a little tattooed. I believe that tattooing in these islands is a religious custom; at least they refused it to several of our gentlemen at Otdia, assuring them that it could only be done in Eregup. In an excursion which I made I have been confirmed in my opinion that this is one of the finest islands; nothing is seen here except fruit-trees and taro plantations. We were told that the island which we had seen in the south belonged to the group of Aur; Kawen and Aur are therefore only ten miles distant from each other.

The 20th of February. The Rurick was surrounded by canoes from morning till evening, and filled with inquiring savages; in the afternoon the princess appeared, to whom I sent several presents into her boat, as she would not venture on board. In the west a large boat was seen, on board of which were twenty-two people, men and women; the utensils of all kinds which they had with them made us suppose that they were on a long voyage. When it came up, the chief of the island of Kawen, Labeleoa, a man of about seven feet high, came on board, and presented me with a roll of *mogan*. He talked a great deal; and, among other things, he advised us to sail to Aur, where the *Tamon Ellip* (great chief) resided. A great many people had assembled on deck, who, conscious of their superior numbers, behaved rather boldly. We were often obliged to put a stop to the impudence of our guests, who were only deterred by the idea that we were supernatural beings, from making themselves masters, by force, of every thing. It was already dark, and the boat which we had sent for water had not yet returned, when the subaltern officer called from shore, that he missed a sailor. As the savages never appeared armed, I had always sent my people unarmed on shore not to create distrust, for which I now bitterly reproached myself. An armed boat was immediately sent on shore, and I at the same time fired a cannon and threw up a sky-rocket; and this phenomenon, so terrible to the savages, produced the desired effect. Scarcely had the cannon been fired, when a dreadful howling arose in the whole island, which lasted for above a quarter of an hour; in the mean time our boats returned: it was quite dark. The sailor who had been missed candidly confessed that love had led him astray, that the girl would not make him happy till after sun-set, and had till then conducted him to a hut in the interior of the island; here a number of islanders had assembled, who would not let him go; they lighted a fire, and stripped him: all fell on the ground as if struck by lightning when the shot was fired, and my sailor fortunately escaped.

The 21st of February. The fright of last night still operated to-day, so that nobody ventured on board till some of our gentlemen had gone on shore. They had asked a great many questions, what the report and flash were? And when they were told that I had on that occasion paid a visit to heaven, my consideration was doubled among them, and they behaved with great propriety. As I intended to leave Airick to-morrow, I paid some visits to-day, and was treated with the most distinguished respect. I could not see the old queen, as admittance was refused me by two sentinels armed with lances. To the princess, however, and some of the nobles I was allowed to make as many presents as I liked. Labeleoa prepared a farewell, *Eb;* three men and three women seated themselves in a half circle; two drummers sat opposite to them, and, with terrible voices, sang the words, *Totabu, Aidara, Möll!* and every motion had a reference to me.

I saw the heron running tame about the huts, and in a wild state along the shore; besides this, there are here only the sandpiper and a species of pigeons. The rats are so bold, that while the savages are eating they come and help themselves.

We found the longitude of our anchoring-place, after an observation between the sun and moon, from fifty distances -	188°	52′	7″	W.
The chronometers gave - -	188	49	25	W.
Longitude from the mean of three observations - -	8	31	11	N.
Variation of the magnetic needle	11	11	0	E.

The observation for the time of high water in the new and full moon gave one hour fifty-two minutes; the greatest difference in the height of the water was four feet. In the direction of N.W. and S.E. the length of the island of Kawen is thirty miles; in breadth eleven miles and a half. The survey of Kawen is founded, like that of Otdia, on points astronomically determined, and I may therefore hope that every navigator who visits that part will be satisfied with the determination of the places. The uniformity of the three groups, Suwarof, Kawen, and Otdia, is probably not accidental; but this structure seems to be peculiar to the corals.

The 22d, at day-break, we weighed anchor, and directed our course to the island of Olot, having promised Langedju to visit him there. Labeleoa, who had intended to accompany us to Aur, sailed off with us, but bent his way to his island of Kawen as soon as he saw us sailing to Olot, where we cast anchor at ten o'clock in the morning, in eight fathoms, over coral sand. We had scarcely arrived when we were visited by Langedju, who was highly delighted to see us; soon after the chief of the island of Torua also came, and I succeeded in reconciling the two rivals with each other, and myself with the latter. On Langedju's invitation, I went on shore with our gentlemen; we found Olot less cultivated than Airick, Tjan, &c.: the population also was less. Langedju conducted me to his taro plantations, where he made a great sacrifice by presenting me with some roots; for though his plantation is the largest in these parts, it would not suffice to support one man four weeks. Near the plantations I perceived a banana tree, which was surrounded with a small fence, and carefully attended to, and seemed to have been but lately transplanted. I learnt from Langedju that the taro as well as the tree had been brought from Aur, and I was pleased at the desire of this people to cultivate their islands to the utmost of their power. It may, perhaps, appear to many of my readers useless to mention such unimportant trifles, but I think by this to furnish the means to those who visit these parts in future, to observe the progress which will certainly take place in the course of time. M. Chamisso also distributed seeds, and taught them the process; out of gratitude for his trouble they stole his knife, but were obliged to surrender it on my insisting very seriously on it. After we had walked about the island for a long time, without meeting any thing worth observation, Langedju conducted us to his hut to entertain us. He (what no one had previously done) sent away, out of mistrust, all his pretty wives, though he, like a perfect courtier, appeared to be my most confidential friend. The meal consisted of sour dough of bread-fruit, of such a very disagreeable taste that we found it difficult to eat any of it. Our host affirmed that he knew Rarick, Langin, and Lagediack; it is therefore to be supposed that the islanders of the different groups keep up an intercourse with each other. Towards evening I went on board, having previously presented Langedju and Labadeny with some useful utensils.

We found the latitude of the island of Olot - -	8° 46′ 4″ N.
Longitude, according to the chronometers, -	188 50 18 W.

The whole group of Kawen consists of sixty-four islands.

The 23d of February we left the island of Olot, in fine weather and a high wind from E.N.E., and directed our course to the passage through which we had penetrated the group. According to the assertion of the islanders, there is a broad passage between the reefs west of Airick, which I have laid down on the chart as it was pointed out to me from the ship. At nine o'clock we had penetrated through the passage by Kawen, and took a S.E. course, at a small distance from the island chain, which forms the lee-side of the group. I called the group of Kawen after our meritorious general, Araktschejef. On account of the high wind we were obliged to carry the top-sails with one reef. At noon, according to observation, we were in latitude 8° 35′ 40″; longitude, according to the chronometers, 189° 3′ 40″. At the same time the sailor on the mast-head discovered in the S.E. an island belonging to the group of Aur, which forms its N.W. point; it is very large, and is called by the savages Pigen. At two o'clock, we already had it to the east, were under the lee in calm water, and proceeded at a small distance along the reef to discover a passage. We had scarcely sailed a mile when we found one about fifty fathoms broad, but so inconvenient, that it could not be passed without danger; notwithstanding this, our desire to examine it prevailed over every fear; the wind had abated a little, the weather was favourable, we spread all our sails, and glided through it. We avoided, by skilful steering, some coral banks which we had not previously remarked, and had soon a view of the whole group, which seemed to us the smallest we had seen in these parts. We approached the island which forms the S.E. point of the group, and is called Aur, and cast anchor at five o'clock in the afternoon under its protection. The lead was often thrown in the middle of the group; we found the depth from twenty-three to twenty-five fathoms, over a bottom of living coral; at our anchoring-place the depth was eighteen fathoms, though we were distant from shore fifty fathoms. We had scarcely anchored when four large boats put off from Aur, and, approaching us within fifty fathoms, quietly remained in one situation, gazing with astonishment at our ship. We addressed them in their language; immediately their fear vanished, and some even ventured on board, where their surprise equalled that of the other savages. Our acquaintances in Otdia and Kawen, whom we mentioned to them by name, contributed to gain their confidence; and two chiefs, who seemed to be very distinguished, and were much tattooed, on hearing this news, ventured to come on board. Upon the whole, we have found the inhabitants of Kawen more importunate than these. We remarked two savages who were tattooed differently from the rest, and, as M. Chamisso observed, spoke a different language. We enquired whether they were natives of that island? They answered, no;

and related to us a long story in their own language, but of which we unfortunately did not understand a word. One of the strangers, a man of about thirty, of a middle size, and an agreeable countenance, pleased me very much; I gave him, after I had made presents to the chiefs, some pieces of iron, which he received with gratitude, though not with the same joy as the other savages. He kept particularly close to me. When the sun was setting, and our guests about to leave us, he took me aside, and, to my great astonishment, expressed a wish to remain with me and never to leave me. I could not suppose that it would last longer than a day, was surprised at the attachment which he had immediately formed for me, and kept him, as the circumstance diverted us all very much. Kadu had scarcely obtained permission, when he turned quickly to his comrades, who were waiting for him, declared to them his intention of remaining on board the ship, and distributed his iron among the chiefs. The astonishment in the boats was beyond description; they tried in vain to shake his resolution; he was immoveable. At last his friend Edock came back, spoke long and seriously to him, and when he found that his persuasion was of no avail, he attempted to drag him by force; but Kadu now used the right of the strongest, he pushed his friend from him, and the boats sailed off. His resolution being inexplicable to me, I conceived a notion that he perhaps intended to steal during the night, and privately to leave the ship, and therefore had the night watch doubled, and his bed made up close to mine on the deck, where I slept, on account of the heat. Kadu felt greatly honoured to sleep close to the Tamon of the ship; he spoke little, whatever pains we took to amuse him, ate every thing that was offered him, and quietly laid himself down to repose. I will give the reader, in a connected detail, the many things which he told of his fate at different times.

Kadu was born in the island of Ulle, belonging to the Carolinas, which must lie at least 1500 English miles to the west from here, and is known only by name on the chart, because Father Cantara, in 1733, was sent from the Ladrones, as missionary to the Carolinas. Kadu left Ulle with Edock, and two other savages, in a boat contrived for sailing, with the intention of fishing at a distant island; a violent storm drove these unfortunate men quite out of their course: they drifted about the sea for eight months, finding, but seldom, fish for their food, and at last landed, in the most pitiable situation, on the island of Aur. The most remarkable part of this voyage is, that it was accomplished against the N.E. monsoon, and must be particularly interesting to those who have been hitherto of opinion that the population of the South Sea Islands commenced from west to east. According to Kadu's account, they had their sail spread during their whole voyage, when the wind permitted, and they plied against the N.E. monsoon, thinking they were under the lee of their island; this may account for their at last coming to Aur. They kept their reckoning by the moon, making a knot in a cord, destined for the purpose, at every new moon. As the sea produced abundance of fish, and they were perfectly acquainted with the art of fishing, they suffered less hunger than thirst, for though they did not neglect during every rain to collect a small stock, they were often totally destitute of fresh water. Kadu, who was the best diver, frequently went down to the bottom of the sea, where it is well known that the water is not so salt, with a cocoa-nut, with only a small opening; but even, if this satisfied the want of the moment, it probably contributed to weaken them. When they perceived the island of Aur, the sight of land did not rejoice them, because every feeling had died within them. Their sails had long been destroyed, their canoe the sport of the winds and the waves, and they patiently expected death, when the inhabitants of Aur sent several canoes to their assistance, and carried them senseless on shore. A Tamon was present at the moment; the iron utensils which the unfortunate men still possessed dazzled their deliverers, and they were on the point of striking the fatal blow, to divide their spoil, when Tigedien, the Tamon of the island of Aur, fortunately came in time to save their lives. When Kadu afterwards offered all his treasures to the preserver of his life, he was generous enough to refuse them; he took only a trifle, and forbade his people, on pain of death, to do any harm to the poor strangers. Kadu, with his companions went to Tigedien's house, who took paternal care of him, and conceived a particular affection for him, on account of his natural understanding and kind heart. According to his reckoning, it must be about three or four years since his arrival here. Kadu was engaged in the woods, when the Rurick came in sight, and he was speedily sent for, as they expected from him, who had travelled far, and was generally accounted a very sensible man, an explanation of this strange phenomenon. He had often told them of ships, which, though they had visited Ulle during his absence, he had heard of; he even knew the names of two men, Lewis and Marmol, who had come from the great island of Britannia; and he, therefore, by the description, knew our ship. Being very partial to the whites, he urged the islanders to go on board, which they, at first, declined, for, according to tradition, the white men devoured the black. How they came to this opinion was an enigma to us, for, except an ancient tradition, that at a very remote period, a large ship had sailed past Kawen, they had no other idea of European ships, but such as had been communicated to them by Kadu. His promise to barter some iron for them, at last induced them to come on board, and here he immediately remained with us, as the reader is already informed. The precaution with which we had him watched was quite superfluous; he slept quietly during the night, and awoke with the first dawn of morning, cheerful and happy.

The 24th of February. We had yesterday informed the islanders that we intended to sail to-day to the island of Stobual, which is eight miles distant from Aur, and which forms the N.E. point, because the sharp-pointed coral bottom was dangerous to the cables. We found the population on the eastern side of the chain of islands to be more considerable in comparison with the other groups: we did not meet with any coral banks, and at ten o'clock reached the island of Stobual, near which we cast anchor in eight fathoms, on a bottom of fine coral sand. It afforded a most pleasing prospect, and, to judge from the number of boats and huts, must be very populous. The group of Kawen was visible from the mast-head. Five boats, which followed us from Aur, and in which there were three Tamons, Tiuraur, Lebeuliet, and Kadu's benefactor, Tigedien, now came on board. Kadu, who had been presented with a yellow cloak, and red apron, walked proudly in his ludicrous finery, without condescending to notice his companions, who gazed on him with astonishment from their boats, and could not conceive the metamorphosis. In vain they cried "Kadu! Kadu!" He did not deign them a look, but walked proudly about on the deck, always taking care to turn himself in such a manner that they might be able to admire his finery. When I learnt that there were three Tamons in the boats, I commissioned Kadu to invite them, as I could not extend the permission to all the savages on account of their numbers; he felt greatly honoured, conducted himself with much dignity, and, after a short speech, first introduced to me Tigedien on deck, as the most distinguished. This old man, with silver-white hair and beard, had a venerable and pleasing countenance, but his tall, strong body was bent with age. He presented me with some rolls of mogan; and, while I was conversing with him, Kadu invited the other chiefs, who were likewise very old, on board. The dress of the Tamons differed but little from that of the other savages; they were only more tattooed, and wore round their necks ornaments of fish-bones, which I afterwards learnt supplied the place of orders. Kadu, to give himself consequence, conducted the guests about the ship, gave them explanations of all the wondrous things which they saw, and knew how to conduct himself so cunningly as to make it appear that he had a perfect idea of every thing he tried to explain; he talked with particular diffuseness on trifling subjects, and generally produced laughter. When they saw a sailor take a pinch of snuff, and questioned him, who had never seen it himself, he was not at all embarrassed; he took up the box, and certainly told them many surprising things respecting it, as they listened to him with the greatest attention; but, when, to make the matter quite plain to them, he took up the snuff to his nose, he threw the box from him, and began to sneeze, and to cry so immoderately, that his astonished auditors ran from him in different directions; but he soon collected himself, and knew how to turn the affair into a joke. Kadu's explanation of the cannon convinced us that he was acquainted with them; for he told them that if the islanders ventured to steal any thing, they would beat down all the cocoa and bread-fruit trees with them; and further related, that Lewis and Marmol, in their visit to Ulle, when the inhabitants had stolen something from the ship, had not ceased shooting down the trees, till the property stolen had been returned. Setting aside this little difference, they must have conducted themselves with much humanity, as Kadu had a very great respect for white men, and liked so much to be with us. The Tamons now attempted to dissuade him from his resolution, but he only shook his head, embraced me and said: I remain with you wherever you go!

We learnt that there was still another chief of the name of Lamary, under whose power the island-groups from Aur to Bigar were subjected, and who was now absent to assemble a military force, with which he intended to seize upon the group of Mediuro, lying to the south of Aur: its inhabitants often make incursions upon Aur, Kawen and Otdia, to seize provisions, of which they are in great want, on account of the numerous population. An incursion on Lamary's island, by which a man lost his life, was now to be punished. Kadu told us that the most shameful pillage was committed upon Otdia; the enemy destroyed every thing they could not carry off: by this information the riddle was solved, why we every where had found newly-planted trees. The people appeared to us unfit for war, and their short, miserable lances confirmed us in this opinion. We now learnt that even the women take a part in the war, loaded with baskets filled with stones, which they throw, as they form the rear-guard, over the heads of their warriors, into the hostile army; they likewise afforded succour to the wounded, and Kadu, who has been in many such battles, assured us that the women were of great service in war. Tigedien, the most distinguished of the three chiefs, supplied the place of Lamary during his absence, and was treated by the people with extraordinary respect. Lebeuliet, the second in rank and dignity, is possessor of the group of Kawen, but resides, in time of peace, in Airick, and the young chief there, as well as the amiable princess, are his children. Tiuraur, the youngest of them, possesses the group of Otdia, and is father to our old friend

Rarick; and it afforded him great pleasure that we were able to give him some account of him. The Tamon returned to Stobual, with many presents, whither they also invited me; but as I had still to make observations to determine the situation of this place, I deferred my visit on shore. Kadu wished to accompany the Tamons, which I permitted him to do, though I was firmly convinced, that inconstant and fickle as the South Sea islanders are, he would not return. He was carried off in triumph. All the canoes followed that of Tigedien, where, elevated to the rank of a distinguished man by our favour, he occupied the place of the Tamon. In the afternoon I went on shore, and immediately took an excursion, accompanied by the active Tamon Tiuraur. The island of Stobual is half a mile in length, and a quarter of a mile in breadth; the fine mould forms here already considerable hills. The palm and bread-fruit trees thrive extraordinarily, and I was agreeably surprised by a young plantation of twenty banana trees. There is more taro here than on the other islands: they. daily brought us some of it. That the root, compared with that of the Sandwich islands, is very small, is probably owing to the want of moisture, though the people assured me that they would thrive very well, if they were not so often destroyed by the inhabitants of Mediuro. Very numerous habitations convinced me of the thick population of this island. In my promenade I came up to the habitation of Lebeuliet, the chief, where a considerable number of men and women formed a circle round Kadu, who had been attracted by his new costume; but I was astonished when I saw him make a speech, at which his audience almost melted into tears: one old woman sobbed aloud. Tigedien's eyes were bathed in tears, and it was easy to observe the effort which it cost Kadu himself to suppress his emotion. He frequently mentioned Aur, Ulle, and Totabu. I was not sufficiently master of the language to understand the connection of the speech, but my supposition seemed correct, that he was taking leave of the chief and the people. As much as I could understand from it, he first spoke of his sufferings on his voyage from Ulle to Aur, painted the generous reception of Tigedien, and concluded with the hope that he might, one day, through me, see his native home again. When Tigedien now began to speak, Kadu shed a flood of tears, the people were deeply moved, and an affectionate embrace of Tigedien and Kadu closed this truly affecting scene; Kadu accompanied us on board, and as his determination to remain with us appeared to be immovable, he was received into the cabin among the officers, which flattered him very much, as he easily perceived the difference between us and the sailors, and thought he belonged to the Tamon of the ship. He sat with us at table, accustomed nimself with incredible readiness to the use of knives and forks, and, in fact, conducted himself with as much propriety and good manners as if he had long associated with civilized people. Our gentlemen treated him with so much kindness that he soon became very much attached to them, and they likewise were happy to have him about them, on account of his good qualities. I cherished the hope that when we had learnt better to understand each other, I should obtain from him much information, as well respecting the Carolinas, as the newly-discovered groups of islands.

The 26th of February, the whole day the Rurick was surrounded by savages who were acquainted with our intention of leaving them to-morrow, and bartered a great number of cocoa-nuts for iron. In the afternoon the Tamons brought us considerable presents of mogan and cocoa-nuts. They regretted our departure exceedingly, and learnt, on their inquiry, that I was going to visit their great Tamon in Ailu and Udirick, as was in fact my intention. As soon as I was alone with my guests in the cabin, they examined very carefully whether any body could overhear us; they entreated, with an air of mystery, but very earnestly, that I would remain here till their military force was assembled, to kill with them all the inhabitants of Mediuro, and then, laden with cocoa-nuts and bread-fruit, return to Aur; they would give me for it an *Eb* every day. This proof of their confidence astonished me, but willingly as I would have defended these poor islanders against their foes, and perhaps have even secured them from future attacks, merely by my appearance, the approaching spring would not allow me time. My refusal dejected them very much; but, to assist them as far as lay in my power, I made them a present of some lances and grappling hooks, which made them inexpressibly happy. Every thing was immediately shown to the people in the boats, who unanimously set up a loud *O—h!*. Tiuraur danced and sang to it a war-song, showing us how he would throw his enemies down; the people roared for joy, and had their foes appeared at that moment, these valiant men, inspired by martial ardour, would certainly have gained the victory. Highly delighted, the Tamons returned on shore. Edock, the friend and fellow-sufferer of Kadu, remained behind to make a last effort to induce him to relent; but all was in vain: all we had given him during his stay with us he gave to his friend, and wept very much on parting with him, when he tore himself from him after a long conversation. Kadu's resolution became to us more inexplicable, when we saw how much he suffered on parting. We are indebted to Edock for a remarkable chart of the geography of the Carolinas, which he had visited, and with which he was better acquainted than his friend. I sketched it according to his statement, and lay it before my readers. Incorrect as it may be, it will perhaps be of some service to future navigators, who intend to examine the Carolinas. The reader knows, from Lagediack's statement, how I discovered three groups. The spaces between the island groups, which, according to Edock's statement, are of the same nature as those examined by ourselves, I have marked with figures, which determine the days' voyages from one group to the other: the single islands are only distinguished by a circle. Sehtiu, Feis, Pelli, and Iap, are said to consist of high, and the others of low land. Iap and Pelli, which are probably the Pelew islands of Wilson, are found marked in all charts. I shall say nothing farther on this subject, as M. Von Chamisso, in his love for the sciences, undertook the trouble of interrogating Kadu, as well on the geography, as on the customs of the Carolinas, and the groups discovered by us. His account, which he furnished for my voyage will be found entertaining and instructive. Edock pointed to the east when I asked him in what direction Ulle lay. His error proves, that an easterly wind had carried the unfortunate men from their island, and that they were driven back to the E. by a S.W. monsoon, which prevailed there; but as they always fancied themselves west of Ulle, they still plied, when they reached the monsoon, always to the east. Edock told us, that on the island-group of Arno, a day's voyage from here to S. W., five men had landed in a boat. They were natives of the group of Lamureck, which lies at a small distance from

Ulle, and is likewise marked on the chart of Father Cantara, among the chain of the Carolinas. A short time ago, when the natives of Arno ventured to make an incursion here, Kadu and Edock recognized the islanders of Lamureck, with whom they had formerly been on friendly terms. As they had heard that the latter had perished, the joy at their unexpected meeting was so great, that Kadu told us they no longer troubled themselves about the battle, but selected a lonely retreat, where they mutually told each other their several misfortunes. The names of the five natives of Lamureck, are Quidal, Pegedu, Uderick, Katulgi, and Udeben. Uderick is the name of an island-group of Radack; and, from the similarity of names, it may easily be concluded, that boats from the Carolinas are often wrecked here. In the afternoon I went on shore to take leave, when I was received by the inhabitants with more kindness than before, as they were now perfectly convinced of my friendship, by the arms which they had received from me. My dog only, which I had brought from Beering's Straits, and which died to-day of convulsions, as the climate did not agree with him, frightened them a little. Before Lebeuliet's residence, a small fleet was equipped, probably against the inhabitants of Mediuro; two of these boats, the largest which I saw here, were thirty-eight feet long. Chamisso passed the night in the Tamon's residence, in the hope of being tattooed, according to his promise; but, as it was not done, we are confirmed in our belief, that tattooing here, is, in some measure, connected with their religion.

In spite of the violent wind, two large boats came from the island of Airick, belonging to the group Kawen, from which it appears that they can sail in a pretty high sea; both of them belonged to the chief Labeleoa, already mentioned, who was much rejoiced to see us again. When, at sunset, the islanders accompanied me to my boat, which had been filled with cocoa-nuts to such a degree, that we could scarcely find room in it, I gave them iron, knives and scissars, and beads to the women, and we parted as the best friends.

I conclude our story with the observations which we made.

Latitude of our anchoring-place, the mean of three observations - - - -	8°	18′	42″	
Longitude, according to the sun and moon - - -	188	48		E.
Longitude, according to the chronometers - - -	188	51	46	
Variation of the compass -	11	58½		E.

The situation of the group Aur is N.W. and S.E.; in this direction, its length is thirteen, its breadth six miles: we counted, in all, thirty-two islands. I called the group after our minister of the marine, Traversey. As our time would not suffer me to examine the islands Arno, Mediuro, and Mille, more closely myself, I could only follow with the compass the direction which the islanders pointed out: the distances are reckoned according to a day's voyage. The population of the group Aur may be estimated between three and four hundred: scanty, when compared to its size; but numerous, compared with the other groups.

The 27th of February. We weighed anchor at day-break, to leave Aur; during the whole night we had heard the drums and songs of the savages: when the sails were set, the noise on shore increased, and Kadu thought that it was done to wish us a happy voyage. At seven o'clock we were already at the passage, by which we had penetrated, and came safely through it. I directed my course northward, to sail close to the island of Kawen; we soon reached it, and, at noon, it lay four miles east of us. We now continued our voyage, sailed round the group of Otdia with a fair wind, and tacked, during the night, under few sails.

The 28th. We had fine weather, and, according to our ship's reckoning, ought to have seen Otdia at seven o'clock in the morning; but as we could descry no land, even from the mast-head, the current must have been the cause of the inaccuracy of our calculation, and which was afterwards proved, as we had been driven sixteen miles to N. W. 11° during the night. To get sight of the group of Otdia, according to which I wished to determine my situation on the chart, I took my course direct to west. In a few hours, land was descried from the mast head to the south. I immediately steered towards it, to discover which island of the group of Otdia we saw, and in three quarters of an hour recognized it to be the island of Ormed, which lay seven miles from us. Thence I steered N.W., in the hope of coming to the group of Ailu. Our latitude at noon, after a good observation, was 9° 51′ 29″; longitude, according to the chronometers, 189° 56′. When we calculated, according to the island of Ormed, we found only a minute's difference, which proved the goodness of the chronometers, as well as the accurate determination of the group of Aur, which gave me more pleasure than any new discovery would have done.

Kadu, to whom we had given a shirt and a light sailor's jacket, was in an excellent humour at this dress, which he liked very much, till the motion of the ship made him sea-sick, and very low-spirited; but his health and good temper soon returned, and he did not appear to miss his friends in the least. At half-past two, three low islands were descried in the N. from the top-mast, which Kadu immediately recognized to be a part of the group of Ailu, where he had once been. He thought that the small island of Temo must lie S.W., and Ligiep farther to the west. We were now under the lee of the group of Ailu, seven miles distant from it, and were, therefore, obliged to tack the whole night to reach it.

The 1st of March, at day-break, we had already gained so much, that we were under the lee at its south point, which is formed by the island of Ailu, after which the whole group is called.

The length of the whole island was scarcely a mile, its breadth a quarter of a mile: it had a pleasing appearance, and was distinguished from the rest by its high palm-trees. We approached the island of Ailu, from which we saw columns of smoke ascending and people walking about. After we had sailed round it, we went along the southern side of the group, which consists wholly of coral reefs; and when we had sailed round this also, we were in calm water: we now continued our course northward, near the reef, in the hope of finding a passage. We soon saw three boats coming through the reefs, but found the passage too narrow for us to penetrate through it. Two boats came so near to us that we could speak with the islanders, and Kadu's joy at seeing some of his old acquaintance, was as great as their astonishment to find him with us. None of them ventured on board, but a long conversation took place from the boats with Kadu, who related to them all he knew about us, and also that he was determined to go with us, but that we intended to pass some days near their island. At this last news the savages expressed much pleasure, showed us, towards the north, a passage, which, according to their opinion, would be wide enough, and we immediately spread more sail to reach it before the evening. We soon found three channels, of which two were deep enough, but only four fathoms wide; the third was from fifty to sixty fathoms broad, but as it was late, and the attempt hazardous, the monsoon also blowing from the pretty narrow opening, I deferred the examination till the next day. We had now overlooked the whole group, the length of which was fifteen, and its breadth five miles; its eastern side was formed by a chain of islands, but the western side consisted of a coral reef.

The 2d of March. — The current had carried us seven miles to the west during the night; we, however, reached the passage. At eight o'clock it appeared almost impossible to penetrate, as it was narrow, and the wind against us; but I hoped that the current, which was setting in, would favour our undertaking, and therefore dispatched Lieutenant Schischmareff to examine the channel. He soon returned with the welcome tidings, that it was indeed only fifty fathoms broad, but deep enough, and free from danger, because the reefs at the entrance, resembling a wall, rose perpendicularly from the bottom. I immediately ordered all the sails to be set to give the Rurick the greatest possible velocity, that in case it should be necessary to turn her against the wind during the passage, she might have force to sail on till every danger was over. We should not have succeeded in this undertaking had we not been favoured by the current: it was very dangerous, and we were greatly rejoiced when we got through, without having come too near the reef, which forms the southern entrance. A species of mackarel had caught itself on the fishing-hook hanging behind the ship, and so the islands paid us a tribute the moment we entered their group. The wind would have carried us in a direct course to the island of Ailu if we had not been intercepted by numerous coral banks, to avoid which, or to sail round them, cost us much time. We had not met with so many shoals in any group; we however found a convenient anchoring-place at noon near Ailu. Three boats immediately came up to our ship, and Kadu, in his sailor's dress, did not neglect to place himself on the deck in such a position that he could be distinctly seen. He condescendingly called out to them, that he was Kadu, they need not fear to come on board; but they, scarcely trusting their eyes, did not venture till after they had had a long conversation with him. After they had sufficiently examined and admired the dress of their old friend, he explained to them with much dignity all the other objects, and thought it quite natural that they should behave to him with as much submission as if he had been a distinguished Tamon. Afterwards he had even the politeness to accompany them on shore, and took, without ceremony, the place of honour in the canoe; the simple savages sung and rejoiced, and carried him on their shoulders through the water without considering that he had only been a common man like themselves a few days before: a zeal which he probably heightened by some old nails which he took with him from the ship to give to them. When he arrived on shore he sat down with much gravity; they all surrounded him, standing, and he related to them his important adventures and experience. In the afternoon we went on shore, where we made the observation that this group must have been produced much later than Otdia, Kawen, &c.; it had but little mould, and was very poor in fruits, considering its size: we did not see the bread-fruit tree at all. The pandanus, which is in abundance on the other islands, was here reared with the greatest care. Fowls, which are tied with strings to the huts, as dogs are with us, are here more frequent: their flesh however does not serve for food, but their feathers for ornament. The long feathers of the tropical birds belong only to their dress of ceremony, but are very seldom met with. We learnt that Lamary had shortly sailed from here to Udirick to assemble a military force, and that Langemui, who had remained here as chief of this group, resided now at the northern point of the island of Capeniur. According to the statement of the savages, the group of Udirick lay north from this place, at the distance of a day's voyage; and I did not doubt a moment that it was the islands of Kutusoff and Suwaroff, discovered by us last year, which, to judge from our longitude, must lie thereabouts. Kadu went with us on board; the islanders accompanied us in their canoes, filled with cocoa-nuts, which they offered us, without desiring any thing in return. On account of the scarcity of fruit among them, I was much moved at this generosity and disinterestedness, and richly rewarded them with iron.

After a good observation, we found the latitude of our anchoring-place 10° 13′ 7″; longitude, according to the chronometers, 188° 58′ 33″.

On the 4th of March, at day-break, the sails were hoisted, and we took our course north, along the chain to the island of Capeniur, which being detained by many coral banks, we did not reach till nine P.M. evening. We lay distant from it fifty fathoms, protected against the wind; and in this convenient anchorage I resolved to remain for several days, to have the sails and cordage repaired, as the Rurick would soon be obliged to go into the tempestuous ocean, where, at this season of the year, we had to expect many storms. The greatest depth in the whole group was twenty fathoms: in our anchoring-place we had only six fathoms; the bottom consisted of white clay; a circumstance which we had not met with in any other group. We had scarcely cast anchor, when we received a visit from Langemui, who had been informed yesterday of Kadu's arrival, and had, on that account, reposed such confidence in us, that he came on board without any fear, to lay some cocoa nuts at my feet. He was a venerable old man, at least eighty, lean, and with grey hair, but a very youthful, lively spirit. I took a great fancy to him, and made him many presents: he invited me to his island, our bond of friendship was sealed, and he soon after left me. In the afternoon I paid him a visit in return, was joyfully received in his house in the bosom of his family, and Kadu, whom I had taken with me, was obliged to relate a great deal respecting our ship. I observed several scars on Langemui's arm, and asked him on what occasion he had received them. He pointed to the W., and while he told me that a long time ago he had visited the inhabitants of Ralick, who had given him these wounds, he fell into such a passion, that he seized his lance, and threw it with such violence against a tree fifteen paces distant, that it remained sticking in it, at the same time crying *Mani mamuan Ralick* (kill man of Ralick). I admired the strength and dexterity with which this old man could still wield the lance, which, however little I had hitherto esteemed it, can inflict a mortal wound at the distance of twenty paces. After I had pacified my host a little, I tried to learn, with Kadu's assistance, who was now accustomed to my pronunciation, what he meant by Ralick ; and received the following information. The chain already known to us from Bigar in the north, to Mille in the south, is called by its inhabitants Radack, as I shall also call it in the sequel; to the W. from the chain of Radack, another chain of islands runs parallel with it, consisting of nine large groups, and three single islands, with a numerous population, and is called Ralick. Langemui explained the whole to me, marking on a mat spread out, with the assistance of small stones, the chain of Radack, which takes its direction from N. to S. as far as to Eregup, and then to S. W. As the groups, as far as we were acquainted with them, were accurately laid down; his information respecting the Ralick chain deserves equal credit. After he had several times told us the names of the groups, he directed us the way which we must take from Ailu to reach them, and which he did in the following clear and sensible manner. A small stone in his hand supplied the place of a canoe; he sailed with it from Ailu at sunrise, arrived at noon, by a S. W. course, at the island of Temo, and from thence, without stopping, to the group of Legiep. (When we discovered these in the sequel, we could not but admire his accurate knowledge of those parts.) He began his voyage the next morning from Legiep, took a western direction, was two days and two nights on the way, and then arrived at the group of Cwadelen, belonging to the Ralick chain. In this manner he described the course, as well as the time of day, with accuracy and clearness. At the time he received his wounds, the two chains were carrying on a violent war with each other, but they are now on friendly terms. Kadu, who likewise affirmed that he was acquainted with the Ralick group, told me that the Tamon Tiuraur, with whom we were already acquainted, had made a voyage to Ralick, had exchanged names with the chief there, and established friendship between the two island chains. The chain of Ralick has two chiefs: Lagadack-nanait and Labondugin: the former is called *Erud Ellip*, (great chief.) The word Tamon was first introduced here by Kadu, as the chiefs are so called in the Carolinas; here, however, he is called Erud. The inhabitants of both chains of islands are said not to differ either in customs, or language; and the group of Odja is reckoned the most populous and largest of the Ralick chain. Langemui assured me that we could arrive there in one and a half day's voyage from Eregup; this, therefore, would be a distance of sixty miles, as I know from experience, that a day's voyage here seldom exceeds forty miles; from Ralick to Radack it would be longer, as one has to contend against the monsoon. It struck me as something very remarkable, to hear Langemui say, that a very long time ago, a ship with white men had been at Odja, and from whom they had procured iron; and that, on the northern group of Bigini, which likewise belongs to this chain, a large ship had been seen sailing by. From this it appears that the Radack chain is quite unknown, except a few groups; and that Ralick may be considered as the chain of the Mulgrave islands, with which we are likewise unacquainted. The chart of the Ralick chain, which, I hope, will be pretty correct, I drew according to Langemui's information; and have added it to my atlas. Of the island of Capeniur, which is only a quarter of a mile * in circumference, nothing more can be said than of Ailu. It is likewise very far back in vegetation, and the population appeared to me very scanty; but perhaps the most had gone with Lamary to Udirick. I discovered several water-pits; and, having fixed upon one for washing our linen, because we could expect no other opportunity before we reached Oonalashka, I daily sent some sailors on shore for this purpose. The islanders were so obliging as to assist my people, and Kadu, who had his linen under his own charge, would not suffer any body to wash it but himself.

At Capeniur, I visited a chief, who, according to his appearance, must have been far above a hundred years old; snow-white woolly hair covered his head and chin: his lean and shrivelled body scarcely resembled that of a human being, and yet he enjoyed the privilege of these happy islanders; his spirits were cheerful, and his mental faculties

* Probably a German mile (4½ English) is meant.—Trans.

unimpaired. It appeared more and more enigmatical to me, how the population could be so scanty, and yet the health of the people so durable, till Kadu gave me the following reason; on account of the scarcity of provisions, the barbarous and revolting law prevails, that no mother is allowed to bring up more than three children; the rest must be sacrificed. We ourselves experienced the beneficent influence of this climate on the body; as, notwithstanding the want of fresh provisions, we were never better in health.

Up to the 6th of March, we enjoyed the finest weather, and a mild E.N.E. wind. In the night, a calm prevailed, which is very unusual; and, on the 7th, the N.E. monsoon changed, for the first time, during our stay at Radack, to N.W. and W. It rained, at the same time, very violently till sunset. On the following morning, the wind again blew from the N.E., and the sun shone brightly. The barometer is not generally subject to great alterations between the tropics, distant from high land; we likewise observed none, setting aside the daily oscillation, except during the west wind, when it fell four lines.

The 9th and 10th, we could do no work on board, on account of the heavy rain, and therefore hurried to finish on the 11th, while the fine weather permitted us. According to our observations, which were confirmed by Kadu, the E.N.E. wind is the most usual at Radack, It is, however, said, that the wind in the months of September and October, sometimes blows from the S.W., and not seldom rises into a furious hurricane, rooting up the cocoa and bread-fruit trees, desolating the islands on the western point of the group, which, he assured me, were sometimes swallowed up by the waves. The savages look forward with terror to this season, which often destroys the crop of bread-fruit. This fruit is gathered only once a-year, and just at that period; for, though it is seen on the trees all the year round, yet they are the fullest in those dangerous months.

Langemui brought on board to-day a young Tamon, of the island of Miadi, which, according to his opinion, must lie now to the east of us, and was, without doubt, the New-year's Island discovered by us; for, according to our calculation, it lay fifty-six miles to the east of us. The young chief made this visit much against his inclination. In a little vessel fitted out for fishing, in which he was alone, he had been surprised in a storm, carried away from his island, and in a few days thrown upon this group. This high-spirited young man was very much tattooed over his whole body, consequently of very high rank; his behaviour was modest, and his inquisitiveness was almost boundless. On my asking when he intended to return to Miadi, he answered, that he would wait for Lamary's arrival, who was to collect a military force there, and go with him. It is astonishing how the savages can make a course of fifty-six miles against the N.E. monsoon, to a point like Miadi, which they can hardly see at six miles' distance. As they only tack, they are two days and a night on the voyage, without any other means to calculate their course than the stars, which they see only with the naked eyes; a skill which the Europeans do not possess.

When Langemui heard that we intended to leave him to-morrow, he was sincerely sorry, and immediately sent out some of his people to gather cocoa-nuts, and others to fish; during the whole night we saw people walking about the reefs with fire, with which they entice the fish, and then harpoon them.

The 13th of March. Already, at day-break, our old friend, Langemui, came to us, loaded with cocoa-nuts and fish, and soon after, we weighed anchor, with a fair fresh E.N.E. wind, and fine weather. The old man stood for a long time in his boat, and bid us adieu, waving both his hands. To this group of Ailu I gave the name of the man under whose command I made the first voyage round the world — Krusenstern.

At seven o'clock, we reached a channel lying towards the north, through which I sailed, as the wind favoured us, though we had only thirty fathoms' breadth. From thence I took the course to N. by W., to reach the group of Udirick, which could be no other than Kutusoff's group.

The mean of several observations, gave for the latitude of our anchorage at the island of Capeniur - - -	10°	17′	25″	N.
The mean of several observations between the sun and moon, gave for the longitude	190	00	40	W.
Declination of the magnetic needle - - -	11	15	30	E.

The mean of our observations, gave for the time of high water at the new and full moon, four hours fifty-three minutes. The greatest difference in the heighth of the water was eight feet.

At three o'clock in the afternoon, the island of Udirick was seen to the north from the mast-head. Towards evening, we approached its southern part, and now clearly distinguished the two groups of Kutusoff and Suwaroff, as well as the channel which separates them, through which we had sailed the preceding year. As it began to grow dark, we tacked under few sails.

The 13th. At day-break, we observed that the current had carried us during the night eight miles to the S.W., 40°, and directed our course to the strait, which is formed on the north by Udirick, and on the south by Togai. At eight o'clock we had passed them, and were under the lee of Suwaroff's group, into which I intended to penetrate; but as we could not find any passage broad and deep enough for our ship, I resolved to remain here only one day under sail to speak to Lamary. Four canoes soon appeared with their chief, and were about to repeat the same ceremonies as last year, when they, to their great astonishment, recognized Kadu. Lamary remained only a short time with us on board, because his people were afraid that we might keep him. He was distinguished from the other islanders less by his dress than his tall and robust person. His face indicated much sense, but his right eye, smaller than the left, gave him a sly look. Kadu afterwards told us that Lamary was now about thirty years old, a native of Arno, and came some years ago to Aur; had murdered its chief without any provocation, and usurped the dominion; thence he had gone to Kawen, and continued to proceed farther to the north with his partisans, to Udirick; had every where murdered the most distinguished chiefs, and now ruled with unlimited sway over the whole chain from Radack to Aur. It is remarkable that the island of Sumatra was anciently known to the Arabs under the name of Lamary; from which one might suppose that the population of the Carolinas, as well as of these groups, had its origin in the Philippine islands; and the more so, as these nations resemble each other very much. After I had made Lamary a few presents on his short visit, he took from his neck a curiously-worked fish-bone, which is worn here for distinction, which he did me the honour to present to me, and immediately left the ship; the other islanders would, however, not be deterred from stopping to hear Kadu's wonderful explanations. I learnt from them that Bigini, the most northern group of the Ralick chain, was exactly to the east of us, and this is perhaps the same known from the chart, under the name of Pescadores, and which has been seen only once. I was told that the island of Bigar was to the N.N.E., and the islanders informed me, that Lamary was soon going there to catch turtle, and to lay them up as a provision for the approaching war.

Two of Kadu's fellow-sufferers, whom Lamary had brought to this island, came to us; one of them, a very old man, was particularly beloved by Kadu, and he resolved to take him with him without saying a word to me. The old Carolinian was beside himself for joy; but fell into a violent passion when I refused his request. He abused Kadu, and besought me to leave the latter in his stead; and in vain were all my representations, that he could not endure a voyage in his old age. I would willingly have complied with his unremitting entreaties, if I had not expected his death almost to a certainty. After the islanders had sufficiently admired all the treasures, Kadu asked my permission to accompany them. M. Chamisso also went, to make himself farther acquainted with the island. The old Carolinian was obliged to be taken by force into the boats, as he would absolutely stay;

and they all left us. In a few hours M. Chamisso and Kadu returned on board, accompanied by several canoes filled with cocoa-nuts. They had not been able to land, as it was impossible to penetrate into the basin of the group, on account of the small opening and the contrary wind; and on the outer side they were unable to pass on account of the violence of the breakers through which Kadu and the other savages swam, while M. Chamisso waited his return in the boat. I now again represented to Kadu, that it was the last moment that he had to reflect. I told him that we should never return to Radack; that he could have no hopes of every going to Ulle; and that he had to expect a long and fatiguing voyage. He threw both his arms round me, vowed to remain with me till death, and nothing remained for me except to keep him, and with a firm determination to provide for him as a father. He distributed in haste all his treasures, and we left Udirick.

The 14th of March. After we had endeavoured the whole night to gain the wind to the east, we found by a good meridian observation, which gave for the latitude 11° 50′ 57″, and for the longitude, according to the chronometers, 190° 26′ 32″, that we had been carried, since yesterday, twenty-six miles by the current, direct to the west. We had not advanced at all to the east, but rather lost ground. This violent current I had observed last year; it seems peculiar to this place, without my being able to explain the cause, and it now prevented me from reaching Bigar. We had fine weather all the day, and a moderate E.N.E. wind. We saw several species of sea-fowl, which indicated that some unknown island was near.

The 15th. We have made another fruitless attempt to reach Bigar; the current has again carried us twenty miles to the west; and, seeing the impossibility of reaching this group, I steered for Oonalashka. The islands discovered by the frigate Cornwallis lay in my way, and I resolved to direct my course in such a manner as to be able to see them. According to Kadu's statement, Bigar likewise forms a circle, but which consists for the most part of reefs, and contains only two small islands; a third is said to lie in the middle of the basin, and that all are covered with but a little mould, and overgrown with low bushes. There are little entrances under the lee of the island for boats, which go there to catch turtle and sea-fowl. I should very much have liked to have penetrated to that place, where, to judge from Kadu's conversation, some light might be obtained respecting the religion of the Radackers. The island is inhabited by a blind god and his two sons; but as he has taken the turtles and sea-fowls under his particular protection, the savages have hit upon the stratagem to adopt the names of the two sons as long as they remain in the island, by which the good, blind divinity is happily deceived, and throws no obstacles in their way. On the other hand, they try to gain the two sons while they are on the way thither by songs full of flattery; from this we see that they attribute to their gods human weaknesses, as the Greeks and Romans did. Their first prayer is, that the rain water may not vanish as long as they are there. They also affirm, that sharks devour no men at Bigar.

The 18th of March. The current daily carried us ten or twelve miles to the west, and I was obliged to take quite a northern course, not to miss the Cornwallis islands. Our meridian observations gave for the latitude 13° 45′ 11″, longitude 190° 38′ 45″. The islands therefore could not be far distant, which was also indicated by a number of sea-birds, which, having fluttered round us the whole day, left us in the evening; and Kadu, who did not suppose any land to be near, suddenly cried out, pointing to the birds, "There is certainly land! the birds are flying home to their young ones; and by this we always know how to find an island when we have lost it." From this it appears that La Peyrouse was wrong in affirming the contrary; as even the savages have made this observation, and take the birds for their guide in the evening.

The 19th. After having tacked during the night under few sails, not to miss the islands, we continued, at dawn, our course to the north, and at seven o'clock the sailor at the top-mast-head gave the news, that he saw land direct before us: according to Arrowsmith's chart, it would lie much more to the east of us. At eight o'clock we distinctly saw several islands overgrown with low bushes, visible only from five to six miles' distance, and they may, therefore, become more dangerous to the navigator than the groups lately discovered by us, which are at least covered with high trees, and give timely warning of every danger. At noon we had sailed round the southern point of the little group, and were under the lee of it, in very calm water, from whence we could clearly overlook it. A coral reef here also forms a circle; the eastern side of which consists of nothing but small islands. The extent, as well as the geographical situation of the group on Arrowsmith's chart very little agrees with our observations. The frigate Cornwallis gives the extent of the group from N. to S. at thirty miles; we, on the contrary, found it only thirteen miles and a half: our longitudes also differ. According to our observation, the group lies 191° 0′ 25″, consequently twenty minutes more to the west than is given by the English frigate: the latitudes nearly agree. When we were, at noon, at the southern point of the group, we found the latitude 14° 39′ 29″. We approached under the lee of the reef within a couple of hundred fathoms. I sent Lieutenant Schischmareff, with a boat, to examine whether we could penetrate into the basin: his endeavours were fruitless; the group had no opening. To judge from the light colour of the water, the depth in the basin cannot be considerable, and, probably, the whole group will soon become one island. An enormous shark swallowed an iron hook as thick as a finger, but was so large and heavy that the hook broke exactly in the middle as we were pulling it up. We soon left the islands, and steered as the wind permitted us, to the north, to reach the 30th degree of latitude, in the same point where we had indications of land the year before.

The 21st of March. At noon we were in latitude 17° 56′, longitude 193° 23′; saw, at sunset, the sea-birds taking their course to N. E., probably to the island of Wakers, which must lie in that direction, but which we could not reach, on account of the north wind.

The 23d: in latitude 20° 15′, longitude 195° 5′. After some violent squalls from all points of the compass, accompanied with rain, we suddenly lost the monsoon, which now began to blow from S. E. and S. We were surprised already by an albatros in this latitude.

The 29th: in latitude 31° 39′, longitude 198° 52′. We looked about in vain for land, and I steered direct to Oonalashka. Though it was early in the year, I wished to arrive there in time, to hasten the completion of the Baydares, which were intended for my voyage to the N.

The 1st and 2d of April: in latitude 34° 3′, longitude 194° 8′. We had such a violent current that it carried us, on the 1st, thirty-six miles to S. W. 23°, and on the 2d, to the S. E. 18°, thirty-six miles and three quarters: we had at the same time a high sea from the S., which caused the current that way to abate a good deal. The temperature of the air had decreased from 23° Reaumur, as we were accustomed to find it between the tropics, to 10°; and it appeared to us very cold.

The 3d: in latitude 34° 27′, longitude 193° 47′. We found the current to-day thirty-four miles to S. W. 81°: we had, at the same time, a faint wind, and saw the water ripple on the surface of the sea, which was caused by the current. So violent a current at a distance from land is a very remarkable phenomenon; but there may, perhaps, have been some undiscovered land near. In the morning, a very singular animal swam near our ship, which seemed to move very little. As the sea was calm, I had a boat put out; the sailors came close to it, and twice the harpoon recoiled, as from a wall. The third time it succeeded; the harpoon entered; and we looked on with much impatience while the animal was towed to the Rurick, and was got up with much trouble. Our naturalists said immediately that it was a fish, called, in natural history, the moon-fish (*Tetrodon Mola*). This singular fish consists only, as may be seen from M. Choris' drawing, of a head, with a very small mouth, is six feet long, and of an oval shape. It probably feeds upon the molluscæ, which it sucks in upon the surface of the water: the exterior part of the fish consists of a gristle, covered with a strong skin, which serves him as armour. The flesh, which is between fish and crabs, appeared to us, who had for a long time tasted nothing fresh, quite a dainty, and, as it weighed five poods (180lbs.), it served our whole crew for several days. A dead albatross, which we drew on board, measured, with expanded wings, seven feet. I took advantage of the calm, went out with a Six-thermometer, and obtained the following results.

Temperature of the air, according to Fahrenheit - - -	60° 00′
On the surface of the water - -	58 50
At a depth of 250 fathoms - -	48 50

Transparency of the water, six fathoms.

The 5th: by a good observation, we were today in latitude 35° 35′, longitude 191° 49′. It appeared that the current had carried us in two days fifty-two miles and three quarters to S.W. 34°. We saw a land-bird, and several species of sea-fowl; sailed past two bamboo canes tied together, and other pieces of wood; all indications of the vicinity of land: all our hopes, however, to discover it remained unfulfilled. We harpooned another moon-fish, and observed that its flesh shone, during the night, like phosphorus. We clearly perceived that we were quitting the fine tropical climates as the northern ocean received us with storms, which always prevail in spring in these parts.

The 13th of April was the frightful day which blasted all my fairest hopes: we were then in latitude 44° 30′, and longitude 181° 8′. Already, on the 11th and 12th, there was a violent storm, with hail and snow. In the night of the 12th and 13th a hurricane arose; the waves, which before ran high, rose in immense masses, such as I had never yet seen: the Rurick suffered beyond description. Immediately after midnight the fury of the hurricane rose to such a degree, that it tore the tops of the waves from the sea, and drove them, in the form of a thick rain, over the surface of the ocean. Nobody, who has not witnessed such a scene, can form an adequate idea of it. It seems as if a direful revolution was at that moment destroying the whole stupendous fabric of nature. I had just relieved Lieutenant Schischmareff. Besides myself, there were four sailors on deck, of whom two were holding the helm; the rest of the crew I had, for greater security, sent into the hold. At four o'clock in the morning, I was just looking at the height of a foaming wave, when it suddenly took its direction to the Rurick, and in the same moment threw me down senseless. The violent pain which I felt on recovering was heightened by the melancholy sight of my ship, whose fate would be inevitable, if the hurricane should rage for another hour; for not a corner of it had escaped the ravages of that furious wave. The first thing I saw, was the broken bowsprit; and an idea may be formed of the violence of the water, which at once dashed in pieces a beam of two feet in diameter. This loss was the more important, as the two other masts could not long withstand the tossing of the ship, and deliverance would then be impossible. The gigantic wave broke the leg of one of my sailors; a subaltern officer was thrown into the sea, but saved himself with much presence of mind, by seizing the rope which hung behind the ship; the steering wheel was broken, the two sailors who held it were much hurt, and I myself, thrown violently with my breast against a corner, suffered severe pain, and was obliged to keep my bed for several days. In this dreadful storm I had an opportunity of admiring the undaunted courage of our sailors; no human assistance however could have been of any avail: but fortunately for seamen, the hurricanes are never of long duration. Kadu had been in great terror during the storm; for he expected, as he expressed himself, that the immense white waves would kill the poor ship; otherwise he was quite secure in the officers' cabin, in warm clothes, only his boots were very troublesome to him. M. Chamisso took great pains daily to learn from him more information respecting Radack and the Carolinas, which he found pretty easy, as Kadu soon understood the Russian language, and we, on the other hand, perfected ourselves in his. When we left Radack, Kadu immediately began to keep an account of time, by making a knot every evening on a string; but after we had been a month at sea, he gave up his account, because he was convinced that we were now wandering at random about the sea, as he had done on his voyage from Ulle to Radack. After the hurricane had abated, and the ship had been put in order, as well as could be done, we continued our voyage to Oonalashka, during which we had to contend with many violent storms.

On the 18th of April, we saw the island of Amuchta, and on the 21st we were in great danger between Oonemack and Oonalashka. Circumstances obliged us to come pretty near to the land lying opposite to us, when a sudden storm drove us on the coast, and we could already calculate the moment of our destruction, when the wind unexpectedly veered: a change very frequent near high land.

The high mountains, covered with ice, of which there are a great many here, astonished Kadu beyond measure. He would not believe that it was land, and it is not surprising, that he who had hitherto seen nothing but small, low islands, covered with the loveliest verdure, should not recognize as land, masses of ice, towering into the clouds. I never saw him regard any thing with more astonishment than snow. To satisfy his curiosity, he one day, when it was falling in very large flakes, gave himself the trouble to catch some, and was seized with a shuddering, when it suddenly vanished in his hand; full of mistrust, he looked at all of us, and thought himself transported into a land of enchantment.

On the 24th, with the assistance of a violent S. wind, we penetrated the channel between the islands of Oonalashka and Oonalga, and, after much tacking, reached the harbour in the night, when a very severe storm arose. I would advise nobody to visit this ocean so early in the year, for the storms are frightful.

Early on the 25th we were visited by the agent of the American Company, M. Kriukof, to offer us all the assistance in his power. The baydares, and all the other things which I had ordered for my voyage to the north, were in hand, and in May he expected the interpreters I had desired from the island of Kodiak. As the ship required many repairs, we immediately went to work, by unrigging her, which had not been done during the whole voyage, and was now necessary, because the rigging had become quite useless. Besides this we discovered that the top part of the mast was rotten; the bowsprit, of which we fortunately had the piece that was broken off, was obliged to be skilfully joined, as there was no wood to be had here for a new one. We were likewise forced to content ourselves with a joined fore-mast, inconvenient as it was, as we should be unable to brace the fore-sails so tight as the contrary wind sometimes requires. The copper sheathing was quite gone in some parts; in others the plates still hung on, which impeded the progress of the ship; she was therefore obliged to be careened and unloaded. It would have been impossible to have had all these repairs finished in proper time if M. Kriukof had not given us active assistance. Every thing still looked wintry in the harbour; the mountains were covered with snow, and the thermometer at noon was only three degrees of heat. During our stay we had for the most part bad weather, which hindered us very much in our work.

On the 27th of May, to our great joy, the two interpreters arrived from Kodiak, who affirmed that they were acquainted with the language of the islands lying to the north of Alaska.

On the 31st, I sent the mate, Chramtschenko, with a small baydare, to survey the islands of Akun and Aketan, where he remained some days.

The 4th of June. A dead whale, which had stranded here, put every thing in motion; the Aleutians flocked in crowds, and swarmed about the half-putrefied fish, as flies about honey. We were prevented from going near it by the nauseous smell. By an arrow which still stuck in the flesh, it was immediately known who had killed it, and consequently was the owner. The district in which such a treasure is stranded is entitled to a part of it, and the inhabitants are allowed to eat as much on the spot as they please, which they accordingly do for twenty-four hours together. The owner often has violent quarrels with the feasters, as they do not think of leaving for him the greatest dainties; that is, the most putrefied parts. Among the greatest delicacies of Oonalashka are the webbed feet of a seal, which are tied in a bladder, buried in the ground, and remain there till they are changed into a stinking jelly.

According to the belief of the Aleutians, the human race proceeded from a dog, which fell from heaven on the island of Oonalashka, and there first produced people. To no theft, like that of Prometheus, will they owe their existence; their ancestor fell directly from heaven, even if he was but a dog.

The 29th of June. After the Rurick was repaired, the baydares put in order for the voyage to the north, and fifteen Aleutians who might be serviceable to us in the little boats, had been embarked, we left Oonalashka. I cannot sufficiently praise the readiness of M. Kriukof to assist us in every thing. He has done as much as lay in his power. With respect to provisions, he supplied us daily with fresh fish, and of the few cows which the American Company possesses, he had one killed for us.

The sailor with his broken leg was now able to go about again; but I was worse, because I had constant pain in my breast; the farther we proceeded to the north, the worse effect had the cold air on me; but, I had still, however, hope and courage to be able to execute my undertaking.

Kadu, who found himself very well in Oonalashka, though he did not like the air, was much surprised that he did not see a single tree on the island, and that neither cocoa-nuts nor bread-fruit were to be had. He took a lively interest in all the new objects which he saw; the Aleutian mode of living under ground did not please him at all; he thought it was better in Radack and Ulle, and asked us whether people lived so at St. Petersburg? We gave him such a splendid description of that city, that he was seized with the greatest desire to see it soon. He looked at the large oxen with astonishment and fear; and his joy was without bounds, that the meat which we ate daily on board the ship was the flesh of these animals. We asked him why he was so rejoiced, and he timidly confessed, that he thought we ate men, and that it might one day be his turn. Soon after our departure from Radack, he had been present when a barrel of salt meat was opened; he observed a piece of the ribs; he remembered the warning of his friends, not to go with us, because we ate the blacks; from that moment, the poor fellow regarded himself as ship-provision, and looked forward, with horror, to the moment when we should be in want of food.

OBSERVATIONS IN OONALASHKA.

The mean of several observations, gave for the latitude of the village of Illiuliuk -	53°	52′	25″	N.
Longitude, according to several observations between the sun and moon - - -	166	31	53	W.
Variation of the magnetic needle - - -	19	24	00	E.
Dip of the magnetic needle -	68	45	00	

The mean of our observations gave, for the time of high water, seven hours, thirty minutes. The greatest difference in the height of the water was five feet and a half.

The 30th of June, at five o'clock in the afternoon, the island of Georgia came in sight, at which I resolved to touch, as M. Kriukof had given me an order to receive there several necessary articles. As the approaching night prevented me from going on shore to-day, I tacked near it during the night, under few sails; and, on the 1st of July, at day-break, steered to the northern and very low point of the island. As soon as we had sailed round it, we perceived many habitations; the shore was covered with an enormous number of sea-lions, which made a frightful noise. A baydare, with three men, came up to us, in one of which was the agent of the American Company, who is, at the same time, chief of the island. He readily offered us his assistance, as soon as he had learnt the object of our coming. As the island has neither a good anchoring-place, nor a harbour near it, the Rurick was obliged to remain under sail as long as I and the scientific gentlemen were on shore, partly to execute some business, and partly to gratify our curiosity about the sea-lions. The landing is very inconvenient, and impossible in a high wind. The agent conducted me to his habitation, which is half under ground; a number of magazines, in which the skins of sea-cats and sea-lions are preserved, lay scattered about. This is the only inhabited spot on the whole island: there are only twenty-five men who, with their wives, have been sent here from the Aleutian islands, under the superintendance of three Russian directors, to kill sea-lions and sea-cats for the company. The agent, who is married to an Aleutian woman, treated us in his half-subterraneous dwelling, with a cup of tea, and then went to the shore, where the sea-lions were lying in great numbers, only two hundred fathoms from the house. These, as well as the sea-cats, live on shore during the time of copulation, and are prepared fearlessly to attack every body who approaches them; while, at other times, the sight of men frightens them in to the sea. The scene which we now witnessed, was to us new and pleasing; we approached these animals within twenty paces: the male arrives to the size of an ox, the females are a little smaller. The lions were engaged in a continual warfare about their mates, as they always try to appropriate several, which they can only conquer from their neighbours by their valour. The heroes are known by the number of their females: they often lie from eight to ten close to each other, that their defender may the more easily protect them, and he always goes raging and roaring around them, ready at every instant, in case of an attack, as the number of lions seems to exceed that of the lionesses. They combat so furiously, that blood is seen gushing out; pieces of flesh fly about, and, not seldom, one of them falls down dead; in which case, the victor immediately enters on the rights of the vanquished, and possesses himself of the widowed seraglio. The combat, however, lasts longer, when several attack one hero; for, as soon as he is vanquished, the allies begin to quarrel among themselves, and do not cease till the most valiant has gained the victory. The roaring of these animals is beyond description; it is heard at sea, during a calm, and when the wind blows from the shore, at the distance of six miles. Their smell cannot be long endured. One is always obliged to keep at some distance from the lions, who, though they can walk but slowly on land, on account of their webbed forefeet, sometimes succeed in a leap of ten paces; and what they seize upon is inevitably lost. One of the Aleutians, who ventured too near the lions, had his whole arm bit off. It was likewise the time now that some of the lionesses brought forth: several lay surrounded by their descendants. The young ones are accounted dainties by the Aleutians and resident Russians. As the agent wished to give us some with us on our voyage, a lioness was frightened, and the young driven a little on shore to kill them. The crying of the young lions has much resemblance to the bleating of sheep. We found the flesh well-tasted in the sequel; but it is said not to be eatable after it is a year old.

The sea-cats, which visit this island only in small numbers, and have their principal quarters at St. Paul, lay apart. The male, which is nearly of the form of a sea-lion, and about the size of a lioness, is as large again as the female. The sea-cat has likewise many mates, but they do not combat for the possession of them, but are constantly employed in watching their females, which take advantage of every moment to fly. The sea-cat is often found lying alone, loudly lamenting the loss of his mate. The skin of these animals is much liked in China, and even in Russia large sums are given for it. The American Company has a fixed and considerable revenue from this and the island of St. Paul. Only thirty years ago, sea-otters were here in such abundance, that a man could catch from two to three hundred of them in an hour; but when these animals, which are accounted by the Aleutians as the most cunning, saw themselves so pursued, they suddenly vanished from these parts. At noon, when my business was concluded, we returned on board; I immediately ordered all the sails to be set, and took my course to the island of St. Paul, where I hoped to be able to procure from the agents residing there woollen-stuff for warm clothing for my crew.

The island of Georgia is of a middling height, runs in a straight line, and seems to have been formed by the eruption of a volcano. Though it is more to the north than Oonalashka, the climate of the latter is more severe, owing to its high mountains. The inhabitants affirm that they have seen for two years past, in the N.E., fire ascending during the night, and are of opinion that there must be a volcano in that direction. This volcano cannot stand on the continent, as the distance is too great to see an explosion; therefore, if the Aleutians have seen rightly, there must be an island there.

The 2d of July. Already, at five o'clock in the morning, we got sight of the island of St. Paul, where we had a calm. We were on the southern part near the small Sea-Otter Island, when the agent of the Company, M. Batujef, came up to us in a baydare, and offered us his services. The Rurick had been taken for a Company's ship, as no others are accustomed to come here; and as they must be always unloaded and loaded again with much haste in the open sea, there being here neither harbour nor anchoring-place, the agent had come to us to make the necessary arrangements, and he now went on shore directly to comply with our wishes. At seven o'clock in the evening, when we had reached the southern part of the island, and were five miles from the Company's settlement, a thick fog spread over the land; notwithstanding this, we were visited by several Aleutians in their baydares, whom we rejoiced very much by giving them some brandy and tobacco. The thermometer during the whole day was at only 4° heat.

The 3d of July. A gun-shot, at five o'clock in the morning, announced to the settlement that we were near; and soon after a baydare appeared with twenty rowers, containing the things we required. As our stock of sugar was almost ended, the agent had the goodness to let us have some of his. He likewise repeated what I had already heard in Oonalashka from M. Kriukof, that, from an eminence in this island, in clear weather, land had been seen in the S.W. Though this supposed land might perhaps be only a cloud, I however thought it my duty to go on shore to obtain an accurate statement of the direction. I took with me my azimuth compass, was conducted to the eminence, and soon observed that the compass turned all round; this motion must have been caused by the quantity of iron which this spot contains; for when I changed my place it was still, and they pointed out to me the direction of the supposed land, which I resolved to look for, in S.W. ½ W. The settlement of the Company here is much more considerable than that in Georgia. There are about two hundred Aleutians under four Russian directors, who are likewise sent from Oonalashka, as the island had no native inhabitants. The land is much lower than in Georgia, and we already found many flowers. The shores here are covered with sea-cats, as those of Georgia with sea-lions, of which there are none here. The Company draws its most considerable profits from the first, on which account the agent, who is chief of both islands, resides on this. Kadu, whom I always took with me on shore, was uncommonly delighted at the warfare of these animals, and made such comical gesticulations, sometimes from astonishment and sometimes from fear, that we could not help laughing heartily at him.

At noon we reached the Rurick, left St. Paul with fair weather and east wind, and directed our course to S.W., to discover, if possible, the land descried.

The latitude of the little Sea-Otter Island we found to be -	57° 2′ 17″ N.
The longitude, according to the chronometers - -	170 10 35 W.

The 4th of July, at noon, we were, according to a good observation, in latitude 56° 30′ 32″; longitude, according to the chronometers, 172° 2′ 37″.

The horizon was clear, the weather serene, and a faint wind from north; the island of St. Paul lay sixty miles distant from us, and we looked about in vain for the new island, which, if it really existed, could not have escaped our observation. Till five o'clock I continued the same course, but when we could not find any land there, I steered north, to the eastern point of St. Lawrence Island. The pain in my breast had continued to grow more violent, but I still hoped to be able to execute my design.

The 10th. At five o'clock in the morning the south-easterly part of St. Lawrence Island was seen from the mast-head N. by W. The land showed itself to us in two small hills, and was distant from us twenty miles. At mid-day it lay to the north of us, at the distance of nine miles, and after we had sailed round a promontory which projected from S.E., and had discovered, in a low spot on the shore, some habitations consisting of tents and huts, I steered thither in order to acquaint myself with its inhabitants. At five o'clock we cast anchor, two miles distant from the village, in four fathoms and a half depth, over a stony bottom. When our boats were put into the water, we saw through our telescopes some people, loaded with baggage, fleeing from their habitations into the mountains, and others arming themselves with lances for our arrival. On the landing-place, when we came, we found twenty tall and robust men, who looked at us with fearful friendliness, without stirring. They had much resemblance to the inhabitants on the western point of this island; and as I perceived the fear which our arrival had occasioned, I did not examine their dwellings, but contented myself with putting the philological skill of our interpreters to the test, which really went so far that they understood each other though with difficulty. We, however, learnt so much that they trade with the Tschukutskoi, from whom they obtain tobacco, iron, and glass-beads, in exchange for skins. While we were conversing with them a baydare was drawn along the strand by dogs, which just came from the Tschukutskoi, and the people shewed us some of the things they had obtained there. They call the inhabitants of the continent of America their brethren; and as they have a constant intercourse with them, and their language is also the same, there seems to be no doubt that these people are of American origin. The eastern part of St. Lawrence Island, on which we then were, they called Kealegack, and the western Tschibocka. Their first question to our interpreter was, where we came from, and whether our intention was to kill them? But after we had given them some glass-beads and tobacco, they lost their suspicions. On my question whether the ice had long left their shores, I received the bad news that it had only left within three days. My hope, therefore, of penetrating Beering's

Straits was blasted, as I could not expect that it would be cleared of ice for fourteen days.

Kadu was here made acquainted with a new people, but whom he could not be persuaded to think human beings on account of their fur-clothing. He made me observe the knives in their sleeves, and therefore always had his clasp-knife at hand to avert every danger from me.

The inconvenient anchorage in which the Rurick lay, did not permit us to remain long on shore. We hastened on board, spread the sails, and steered to the northern point of the island. The small island laid down on Cook's chart as a single one, consists, as we perceived in sailing by, of two, separated by a narrow channel. At twelve o'clock at night, when we were about to anchor at the northern promontory, we perceived, to our terror, firm ice, which extended as far as the eye could see to N. E., and then to the N., covering the whole surface of the ocean. My melancholy situation, which had daily grown worse since we had left Oonalashka, received here the last blow. The cold air so affected my lungs, that I lost my breath, and at last spasms in the chest, faintings, and spitting of blood ensued. I now for the first time perceived that my situation was worse than I would hitherto believe; and the physician seriously declared to me that I could not remain near the ice. It cost me a long and severe contest; more than once I resolved to brave death, and accomplish my undertaking; but when I reflected that we had a difficult voyage to our own country still before us, and perhaps the preservation of the Rurick, and the lives of my companions depended on mine, I then felt that I must suppress my ambition. The only thing which supported me in this contest, was the conscientious assurance of having strictly fulfilled my duty. I signified to the crew, in writing, that my ill health obliged me to return to Oonalashka. The moment I signed the paper was the most painful in my life, for with this stroke of the pen, I gave up the ardent and long-cherished wish of my heart.

CHAPTER XIII.

FROM THE ST. LAWRENCE ISLANDS TO GUAHON.

On the way back to Oonalashka, where we arrived on the 22d, nothing remarkable occurred, except that, in sailing, we struck upon a sleeping whale; the ship was so violently shaken that I, who was lying in the cabin in bed, thought that we had run on a shoal. The fish, awakened in a fright, had at once darted to a great distance, and then dived to the bottom of the sea. In Oonalashka we found every thing in the loveliest blossom, which was very beneficial to us all, and particularly relieved the oppression on my chest. M. Kriukof had the kindness to prepare for me a small house in the country, where I recovered pretty tolerably. We here employed our time in baking biscuits of bad flour; for, as our little Rurick could scarcely hold a stock of provision for two years, we had for this six months past been put on half-allowance, and could not, in spite of this economy, make shift for more than three months. At Oonalashka we had taken in, for our voyage to the north, a large quantity of stock-fish, which was in some measure to supply the want of biscuit. The officers' table was served as meagre as that of the sailors, and the only difference was, that the stock-fish was sometimes dressed as a pudding, and at others as a sauce. The dish was considered the most elegant, when treacle was poured over it, though the fourth part consisted of sea-water, which richly imparted to it its salt, bitter taste. We received the treacle from Oonalashka and St. Paul, from the agent, who had it from China, where it is brought by ships from the united states of America. I could not learn who it was that first thought of mixing sea-water with the treacle; but he probably found an advantage in it: we, however, experienced its bad effects, by a swelling of the stomach.

The want of fresh provisions, and the bad state of the Rurick, which was in absolute need of repair, would not allow me to return through Torres' Strait, as my instructions prescribed; I therefore resolved to go to Manilla, where I hoped to find all we stood in need of. To make this voyage useful, I intended to call at the Sandwich islands, to take in plants, and domestic animals, and bring them to the inhabitants of Radack, and thus do a service, not only to them, but to such navigators as may in future visit those islands. While I was at Radack, I proposed to employ some time in examining the Ralick chain, and then take my course to the Ladrones; for in this ocean, strewed with dangerous islands, many new discoveries might be expected.

Before I quit Oonalashka, I will communicate M. Kriukof's account of the island, which rose from the sea, and of which event he was himself an eye-witness.

In the year 1796, on the 7th of May, M. Kriukof had arrived on the northern point of the island of Oomnack, at a small distance to the east of Oonalashka, with several hunters, who had selected it as a retreat after a fatiguing excursion. They intended to continue their voyage to Oonalashka in their large baydares the next day, but were prevented by a violent storm from the N. W., accompanied with rain. This storm lasted till the 8th; upon which the weather became fine, and they saw, to the N., several miles from land, a column of smoke ascending from the sea; towards evening they observed under the smoke something black, which rose but a little above the surface of the water. During the night, fire ascended into the air near the spot, and sometimes so violent, and to such a height, that on their island, which was ten miles distant, every thing could be distinctly seen by its light. An earthquake shook their island, and a frightful noise echoed from the mountains in the S. * The poor hunters were in deadly anxiety; the rising island threw stones towards them, and they every moment expected to perish. At the rising of the sun

* All the Aleutian islands are of volcanic origin, and seem to be the production of a dreadful revolution; nothing is seen but high conical mountains, of which many exceed the Pico of Teneriffe in height: formerly they threw out fire, and some of them still continue to do so.

the quaking of the earth ceased, the fire visibly decreased, and they now plainly saw an island of the form of a pointed black cap. When Kriukof visited the island of Oomnack, a month after, he found the new island, which, during that time, had continued to emit fire considerably higher. After that time it threw out less fire, but more smoke: it had increased in height and circumference, and often changed its form. For four years no more smoke was seen, and in the eighth year, (1804,) the hunters resolved to visit it, as they observed that many sea-lions resorted to it. The water round the island was found warm, and the island itself so hot in many places that they could not tread on it. The island is said to increase in height and extent to this day. A very sensible Russian who was there, told me, that it is two miles and a half in circumference, and was three hundred and fifty feet high: for three miles around, the sea is strewn with stones. He found the island warm from the middle to the summit, and the smoke which ascended from the crater appeared to him to have an agreeable smell. Some hundred fathoms to the north of this island is a rocky pillar of considerable height, mentioned by Cook: he took it, at a distance, to be a ship under sail. Our Russian navigator, Saritschef, has seen this pillar, which has kept its place since time immemorial. Experience has however now taught us that it is connected under water with the island of Oonemack.

The 18th of August, at ten o'clock in the morning, with fine weather and a S. W. wind, I left Oonalashka for the third time. Kadu, on hearing that we were going to Radack, was greatly delighted, and from the moment he learnt it, laid up a collection of old rusty nails and useless pieces of iron; on the shore he collected stones, which appeared to him good for whetting: in short, he did all he was able to be of service to his friends. But he was still fully resolved on no account to stay there: Petersburg seemed to him too attractive. The great quantity of iron which we carried as ballast on board the ship, appeared to him to indicate immense riches: when it was unloaded at Oonalashka, he could scarcely trust his eyes. I left a part of the iron with the company, as they were in great want of it there. I also supplied them with tobacco, which is to them a very important article, as the Aleutians will undertake nothing without chewing this herb. I farther undertook to convey from here to St. Petersburg several hundred Morse-teeth, to prove to the Company, how much I wished to show my gratitude for the kind reception I met with in its colonies.

At noon, we were already out of the bay, which forms the entrance of the harbour. A number of whales surrounded us, which threw themselves high in the air, and then sunk down with a dreadful noise, by which the water was dashed about in foam. It is scarcely credible, that such an enormous, and, to appearance, helpless creature, should be able to rise so high above the surface of the water. The Aleutians count seven species of whales, of which the most are probably unknown in natural history. One of these species is a rapacious animal, which is well known not to be the case with other whales, as they have no teeth, and only live on small fishes. This animal, of the size of the largest whale, is furnished with dreadful jaws, full of large teeth. It devours every thing it can catch, and often pursues the Aleutians, whose little baydares, if it is able to overtake them, it upsets with one blow of its tail. It is said that a baydare, with twenty-four oars, and thirty men, was lately destroyed by the blow of such a monster, near Oonalashka.

The Russians and Aleutians relate, that if a piece of the blubber of this animal is swallowed, it has the property of immediately passing through the body undigested.

M. Kriukof's description of a sea-animal which pursued him at Beering's Island, where he had gone for the purpose of hunting, is very remarkable: several Aleutians affirm they have often seen this animal. It is of the shape of the red serpent, and is immensely long; the head resembles that of a sea-lion, and two disproportionately large eyes give it a frightful appearance. "It was very fortunate for us," said Kriukof, "that we were so near land, or else the monster would have swallowed us: it stretched its head far above the water, looked about for prey, and vanished. The head soon appeared again, and that considerably nearer: we rowed with all our might, and were very happy to have reached the shore before the serpent. The sea-lions were so terrified at the sight, that some rushed into the water, and others hid themselves on the shore. The sea often throws up pieces of flesh, which, according to their opinion, is that of this serpent, which no animal, not even the raven, will touch. Some Aleutians, who had once tasted some of it, suddenly died. If a sea serpent really has been seen on the coast of North America, it may have been one of this frightful species.

The Aleutians also tell stories of a gigantic polypus. It has happened, that a polypus has thrown its long arms, which are twice as thick as a strong man's arm, round the baydare of an Aleutian, and would have carried it into the abyss, if the Aleutian had not had the presence of mind to cut through with his knife the fleshy arm of the polypus, which was furnished with large suckers. The polypus remains with the body fast at the bottom of the sea, and generally chooses a place from which it can reach the surface with its arms. The last accident happened in the passage which is formed by the southern point of the island of Oomnack, and the little island lying near it; no ship can venture in on account of the shoal. This island, which is only five miles long, one mile broad, and very low, is not to be found on any chart; and I hope, that our survey of the Aleutian islands, and from these towards the east, as far as the western coast of the island of Oonemack, is pretty correct. We are much indebted to M. Saritschef, he being the first who made a chart of the Aleutian islands.

When we were out of the bay, we took a north-easterly course, to reach the channel between the islands Oonemack and Akun, which is far the safest passage into the ocean. Near it, on the 19th, in the morning, the wind began to blow so violently from N.E., that it detained us till the 20th in the Strait, which we did not pass till towards evening, when the wind veered to west.

On the 21st, the wind blew high from the east, and at eight o'clock, we clearly saw the two high mountains on Oonemack, and the continent of Alashka: the latter smoked very much. Some years ago this volcano had a violent eruption, in consequence of which its conical-shaped point fell in. The noise caused by this was so dreadful, that in the mountains of Oonalashka, though ten miles distant from it, it resembled thunder. During this explosion, the mountain threw up a number of balls, of the size of a walnut, some of which are in my possession, the chief ingredients in which are lava and iron.

On the 23d, the wind veered to the south, and disappointed my hope of soon reaching the tropics.

A number of albatrosses flew round our ships: the idea of many learned men now struck me, that these birds fly from the north to Cape Horn to build their nests. Common sense contradicts this assertion. The Aleutians are accustomed to search for the nests of the albatrosses on the summits of their mountains, and are very fond of the eggs. On the island of Oonemack and other volcanic islands, the birds build so high that it is difficult for the Aleutians to reach their nests. They shoot them with arrows in autumn, when they are fattest: their fat is esteemed a great delicacy. The black albatrosses, which are by many thought to be the young of the white ones, are asserted, by the Aleutians, to be a distinct species.

The 10th of September. To our great joy the wind changed to-day to the north. At noon we found the latitude 40° 10′, longitude 147° 18′. We had spent eighteen days in constant tacking in a thick fog, or fine rain; and at the same time we had such a violent storm, that we were obliged to have recourse to the storm-sails. As the sun now shone more brightly, we found our longitude, according to the chronometers, five degrees different from the longitude of our ship-reckoning. The current, therefore, had in these eighteen days carried us so far to the east. The heat, which sensibly increased as we advanced quickly towards the south, was very beneficial to my health.

The 13th. When we were in latitude 36° 9′,

longitude 148° 9′, I took advantage of the calm to make an experiment with the Six-thermometer, and obtained the following results:

Temperature of the air on the surface of the ocean			+ 73° 00
Temperature of the water on the surface of the ocean			+ 71 90
Temperature of the water at the depth of	25	fathoms	+ 57 10
- - - - - -	100	-	+ 52 80
- - - - - -	300	-	+ 44 00

The transparency of the water was about 13 fathoms.

In the calm, on the 14th of September, of which I also took advantage, the latitude was 35° 51′, longitude 147° 38′.

Temperature of the air on the surface of the ocean			+ 75° 00
Temperature of the water on the surface of the ocean			+ 72 20
Temperature of the water at the depth of	4	fathoms	+ 72 00
- - - - - - -	8	-	+ 70 90
- - - - - - -	15	-	+ 68 10
- - - - - - -	25	-	+ 57 60
- - - - - - -	50	-	+ 54 00
- - - - - - -	100	-	+ 51 00
- - - - - - -	408	-	+ 42 00

Transparency of the water 11 fathoms.

While I was making these observations in the boat, a shark came so close to us that a sailor struck him with an oar, for which he, however, revenged himself by biting through the string of my Six-thermometer, and I lost the instrument in a moment, when I had, for the first time, sunk it to five hundred fathoms, and was very anxious for the result. The string I had to this was made in Oonalashka of whale-sinews.

The 21st: latitude 27° 58′, longitude 152° 27′. Three small snipes fluttered round the ship, and at last disappeared. Though these birds generally indicate land being near, we looked about for it in vain. The Spaniards are of opinion that there is an island in this neighbourhood, and call it St. Maria-la-gorta.

I opened to-day a tin-box with plumb-pudding which had been prepared in England in 1815, and was still quite excellent.

The 22d: latitude 27° 50′, longitude 152° 22′. A perfect calm allowed me to-day to use my second and last Six-thermometer.

Temperature of the air on the surface of the ocean	+ 77° 10
Temperature of the water on the surface of the ocean	+ 77 00
At the depth of 5 fathoms - - - - - -	+ 75 00
- - 10 - - - - - -	+ 75 00
- - 25 - - - - - -	+ 73 70
- - 50 - - - - - -	+ 67 20
- 100 - - - - - -	+ 61 00
- - 200 - - - - - -	+ 51 50

Transparency of the water 16 fathoms.

The 23d: latitude 26° 41′, longitude 152° 32′. The N. E. monsoon again commenced.

The 26th. At seven o'clock in the morning a mountain in the S. W. came in sight, and I recognized it for Mouna Roa, which lies on the N. E. side of the island of Owhyee. The whole island soon appeared; at noon, the wind being very faint, we were still thirteen miles distant, but at sunset a breeze rose in the N., which slowly brought us to the north point of the island. The moon shone bright, the weather was fine, and I resolved to sail round the point during the night. At midnight we were already under the lee of Owhyhee, within four miles of land: the bright moon, and the many fires of the savages on shore, made our course safe and agreeable.

The 27th. The dawning morn brought with it a dead calm; we were opposite Young's possessions in Tocahai Bay: the sun, rising without a cloud, announced to us a fine day, and Mouna Roa stood before us in cloudless majesty. We were still admiring the beauty of the scene, when suddenly a young and very handsome girl caught our attention. She had taken advantage of the calm to approach us alone in a small boat; expressive gestures explained her object, and she was very angry when she saw herself laughed at. Kadu, who was already delighted at the sight of his dear cocoa-trees, of which he had been so long deprived, was enraptured at the sight of the girl. He tried to make himself understood in all the languages with which he was acquainted, but as she likewise did not understand his Russian conversation, he earnestly entreated me to suffer her to come on board, which I however refused for good reasons. He at last contented himself with throwing to the fair one all his glass-beads, and nodded to her as long as she remained in sight. A second boat, with five Sandwich islanders, restored him to his good temper. The savages brought us taro roots and water-melons, for which they were paid very dear, and we learnt from them that Tamaahmaah was at that time at Owhyee. Towards noon a faint sea-wind arose, by which we proceeded along the coast to the S. I wished to reach Karakakooa Bay, where I hoped to find Tamaahmaah, to conclude with him a bargain for fresh provisions. At sunset we had, however, as is generally the case here, a dead calm, and were still far from our destination.

The 28th, in the morning, a faint land-wind brought us to the low point of land behind which lies Teiatatooa Bay, the same where the king had received me so kindly the preceding year. Two chiefs, who were just sailing out to fish, visited us, and were recognised by us as old acquaintances: they told us that Tamaahmaah was in this bay; and, after they had succeeded in deceiving us, they sailed off much rejoiced. Soon after we saw a second canoe sailing up to us with incredible swiftness. I immediately ordered the ship to lie to, and we saw in the boat our old fellow-traveller, Mr. Elliot de Castro, who had seen the Rurick through a telescope, and hastened to us. As we had already sailed by Tamaahmaah's residence, I had the ship put about, and we steered to the bay in which the king then was, on account of the bonéto fishery. I accepted, with much pleasure, Mr. Elliot's offer to go in his boat, as I should thus be enabled to see the king much sooner, and conclude my business to-day. Our scientific gentlemen, as well as Kadu, also embraced this opportunity, by which we reached, at noon, the king's quarters, which were on the shore on a bed of lava exposed to the scorching sun. His residence the preceding year was much more agreeable than here, where the eyes fell on rugged rocks. About twenty years ago, a very high mountain near this place emitted fire; the lava flowed into the sea, and formed the bed on which the straw tent was erected, and which scarcely held three persons. The king lived as inconveniently as the nobles of his kingdom, whom he has always with him. When they are displeased, he says, with justice, "I am not a straw better off than you; if I let you remain on your estates, you become as fat as your hogs, and have no other thoughts but to injure your king." As he had resided two months in this disagreeable place, and fully put the patience of his chiefs to the test, he intended, in a few days, to select a more agreeable situation; and this resolution he made known to his suite with the following remark: "You will now know better how to appreciate what is agreeable." Tamaahmaah, shortly before our arrival, had gone to the bonèto fishery, and we were in the interim conducted by Elliot to his wives, who were sitting in the middle of a tent, on fine mats, under an umbrella of white sail-cloth, refreshing themselves with water-melons. They were all three greatly rejoiced to see us again. I was obliged to sit next to Kahumanna, who, after having put to us some indifferent questions, ordered water-melons to be brought, which were very cooling in the heat. Her politeness went so far, that she ordered a *Kanaka* to drive away the flies from me, with a tuft of red feathers; she even cut out, herself, the middle of the melon, and put the piece into my mouth with her own royal hands; in doing which the royal nails, three inches long, incommoded me not a little. On this occasion she had me asked, whether the favourite queen in our country was as polite to strangers as she. I answered that we had indeed a very condescending and good queen, but only one. At this Kahumanna was greatly astonished, as she had heard that our king was a great monarch, and therefore, according to her notions, was entitled to have many wives. Kadu was looked at with much curiosity, his long ear-laps struck the queen greatly, and were closely examined by her. The people, who soon learnt that he came from a newly-discovered island, crowded to see him; several chiefs, and even the queens, made him many presents. He at first behaved timidly, but he was much pleased at it, and more particularly when two young girls took hold of him to show him about the tent. The sun had almost set, when the king returned from the bonèto fishing, which is carried on at a distance from shore with lines and hooks. He did not give himself time to dress, but came naked up to me, and welcomed me with a cordial shake of the hand. A minister dragged two bonétos behind him, and the king said, at the same time having one laid at my feet, "I have caught this fish myself, and I beg you to accept it as a token of my friendship." His wardrobe was now fetched, which consisted of a shirt, a pair of old velveteen small-clothes, a red waistcoat, and a black neckcloth; and, without any ceremony, he dressed himself in my presence. His elegantly embroidered uniform he puts on only on very solemn occasions, and even then unwillingly. He once said to Elliot, "The uniforms which King George (so he calls the King of England) wears, shine very much, but can be of no service to me, because Tamaahmaah outshines every thing!" I observed, while he was dressing, many scars on his body, and, on my asking in what campaign he had received them, he answered, pointing to the N. E., "I have conquered these islands; and these scars prove that I deserve to be king of the whole group." After he had dressed himself, he sat down on a mat in the open air, close to his tent; a mat was likewise spread for me, and the nobles of his kingdom stood in a circle round us. A gourd-shell was brought, filled with taro dough, and, while he quickly crammed the dough into his mouth with his fore-finger, he entertained us with the bonéto fishery. Kadu interested the king very much, and he, on his part, treated the king with the greatest respect, whose splendid possessions, in his opinion, rendered him the first tamon in the world As I had no time to lose, immediately after the meal I mentioned to the king the provisions I wished to take in at Woahoo. The king answered, "I can transact no business of any kind with you to-day, because my son Lio-Lio has had this night an ominous dream, portending misfortune. The dog of all dogs, in his dream, swallowed the Queen Kahumanna, and spit her out as a most frightful monster, which immediately began to ravage the country. I must therefore suppose that you are to-day the bringers of misfortune." I assured the king that our ship concealed no monster of the kind, which the dog of all dogs had spit out, and that, on the contrary, he could not have a more sincere friend than myself; and after much persuasion, I succeeded in being dismissed to-day. One of his chiefs, Kareimoku, a relation of the Governor of Woahoo, was obliged to sit on the ground, and receive his commands, which were, that we should have as much provisions as in the last year, and be treated with as much friendship. He then turned to me, saying, "You may now begin your voyage to Woahoo; this chief you take with you, he will provide for all your wants. I demand no payment for the provisions; but if you have some iron which you can spare, I will thank you for it, as I use it in building my ships." I readily promised to send him as much iron as I could spare, and hastened, favoured by a faint land-wind, to commence my voyage to Woahoo. Our companion, the young Kareimoku, conducted himself with much propriety; two *Kanakas*, whom he had with him to attend upon him, proved that he belonged to a distinguished class of the Ierries. A very faint land-wind made our voyage tedious: we were becalmed a whole day near the island of Ranai. One must take care not to get too near under the lee of this island, because the monsoon is intercepted by its height.

The 1st of October. At day-break we saw Woahoo, and at five o'clock at noon anchored at Hana-rara. A brig with the American flag, which we had previously seen sailing from the north, through the channel between Woahoo and Morarai, came close up to us soon after we had anchored. The ship really belonged, as I afterwards learnt, to the nation whose flag it bore. Baranof had hired it in Sitka, to carry a cargo of skins to Okhotzk, and the captain was just then returned after having finished his business. As soon as I had cast anchor I went on shore, where the young Kareimoku had previously gone in a canoe of the natives. We found the harbour very animated; eight ships were lying at anchor, of which six carried the North American, and one Tamaahmaah's flag; the eighth was lying on the strand, and belonged to the Russian American Company.

As I approached the little fleet, the guns in the American ships were fired; a politeness shown to me as commander of a Russian ship of war. I was kindly received on the landing-place by the captains, and accompanied to Kareimoku's habitation, who was greatly rejoiced to see me again. While I was still at a distance he called out to me *Aroha!* (welcome). Three guns were fired from the fortress, and at every shot he repeated *Aroha!* pressing my hand. He told me, through Young, that he had already received Tamaahmaah's commands by his commissioners; but even without these he would have provided every thing out of love for the Rurick. I asked for boats to tow me in, but the American captains begged me to accept their boats instead, which they promised to send early to-morrow.

The 2d. According to the custom prevailing here, a gun was fired at day-break, and immediately the boats appeared, which towed us in, and brought us to anchor in the same place where we had lain the preceding year.

We had scarcely arrived when Kareimoku, accompanied by Young, came on board, followed by a large boat laden with vegetables, fruit, and a hog. Kareimoku felt himself much flattered when I received him on deck with a salute of three guns: the fortress saluted with seven guns, and we answered with the same number. Kareimoku informed me, with much pleasure, that Dr. Scheffer had been expelled by the king, as well as the people, from the island of Atooi, and had landed here some time ago with his crew, consisting of a hundred Aleutians, and several Russians in the ship Kodiak, which lay now on the strand. The ship was in such a miserable condition that the crew, in the voyage from Atooi to Woahoo, had been compelled to pump continually, to prevent her sinking, and, therefore, the fugitives, as soon as they reached this harbour, were obliged to run her a-ground. Kareimoku told me that he had kindly received the unfortunate Aleutians and Russians, as he did not wish to return evil for evil; even Scheffer had been permitted to go on board a ship from the United States, which a few days before our arrival had sailed for Canton. Kareimoku had scarcely finished his account, when M. Taracanof, agent to the Russian American Company, came on board, with several of the Company's officers. Taracanof, who, according to Baranof's orders, was placed entirely under Scheffer's command, expressed his disapprobation of the proceedings in Atooi, through which they had all been placed in the greatest danger of their lives, and he considered it a great miracle that, in their flight from Atooi, only three Aleutians had been shot, as Tamary, who regarded them all as his bitterest enemies, might easily have killed many more. He also mentioned their dangerous voyage hither, and was now, with his people, in the most miserable situation, as of course they could not have provisions without payment. Fortunately I had laid in such a provision of stock-fish at Oonalashka, as enabled me to send these poor creatures provisions for a month. Taracanof, who seemed to me a very sensible man, had concluded a contract with M. Hebet, the owner of two ships lying here, by which he bound himself to maintain and clothe the Aleutians a whole year, on condition that he should bring them to California, where they should employ themselves in catching sea-otters on the islands there. After the expiration of this year, Hebet brings them to Sitka, and gives the Company half of the skins obtained. This contract was advantageous to the Company, which often lets out Aleutians in this manner; for these unhappy people will be the victims of their oppressors, as long as the Company is subjected to the dictates of a monster, who purchases every gain with the blood of his fellow-creatures.

The necessary preparations were now made on board the ship, that we might leave Woahoo as soon as possible. M. Chamisso, in the mean time, made a short tour through the country.

The 6th of October. To-day the American brig Boston ran in here, after having doubled Cape Horn and touched at Sitka, and intended from this place to proceed to Canton. The captain let us have, on payment, some biscuit, of which we were entirely out.

Kadu had made himself many friends here, where several things excited his astonishment; among others, he was once extremely frightened by a man on horseback, as he took him for a dreadful monster. The islanders took pleasure in teaching him, and as he particularly interested himself in the cultivation of the land, I hoped, through him, to instruct the inhabitants of Radack with respect to the plants which I intended to take with me.

The 8th. The captain of an American schooner has concluded with Kareimoku a bargain for a cargo of sanders-wood, giving in return a copper-bottomed vessel. This shows how dear the Americans are paid in China for sanders-wood. Several ships which lay here at anchor paid for this wood in goods or piasters, and it is delivered to them in the presence of Kareimoku, according to weight.

I took a walk every evening when the air was cooled. This may be done here without any fear of attack, for though one frequently meets with drunken people, they are in this state the most cheerful and affectionate. They intoxicate themselves with the Ava-root, which is prepared in the same manner as in the other South Sea islands, with this exception, that here only the old women chew the root, and the young ones only spit into the dish to dilute the pap. How unhealthy the constant use of this root must be is proved by the many ulcers with which the inhabitants are afflicted. The chief persons intoxicate themselves oftener with rum, which they procure from the Americans. Since the Europeans have introduced the use of brandy and tobacco, and brought at the same time many bad disorders, the population has visibly decreased. Many fields also lie uncultivated, as the inhabitants are obliged to fell sanders-wood. On my way to the plantations, I was met by two boys, carrying large bundles of banana, and stopping at every hundred steps to make the people present aware of their coming by a loud cry. The men immediately threw themselves on the ground, covering their faces with both hands, and did not rise till the boys had passed; of the women more was required, for they were obliged to undress on seeing the boys. I was told, this evening, a very important taboo began; the bananas were carried into the morai, where they were offered to the gods, and therefore this submission must be shown to the bearers of the holy fruit. Soon after I went to the house of a distinguished chief, an acquaintance of mine, who was sitting with several others before the door waiting for the sun setting to go to the morai; he saluted me in a friendly manner, but warned me not to touch him, as I should thus become *taboo*, and must go into the morai. During this dangerous time the women are not allowed to be seen by their husbands, and should one commit the crime of touching him, she must atone for it with her life.

The corpse of a woman which I saw floating about in the harbour was pelted by grown-up persons and children with stones. I was informed that the woman had transgressed a taboo.

The 11th of October. I was attracted to-day, by the sound of a muffled drum, to the morai. As it was not *taboo-day*, I supposed that the people engaged in it were priests. The attention with which I looked at them was observed from the morai; two islanders appeared, and saluted me with *Aroha jeri nue!* (hail, great chief!) and proposed to me to enter. I was astonished that this permission was extended to me, and was not without some fear that the priests might take it into their heads to offer me to their gods. Separated from my people, who could not even learn where I had remained, I resolved at least to be on my guard, and was conducted through the sacred gate. As this morai, as I have previously said, was built up in haste, after the destruction of the old one, it could not give me a right idea of such a sanctuary. I found here only a piece of land fifty fathoms square, fenced in all round with bamboo canes; in the middle of the place six small houses standing close to each other, formed a half circle: each of these chapels was surrounded with a low bamboo fence, above which the colossal heads of the gods looked over like sentinels. The necks which supported their enormous heads were decorated with hogs'-flesh; and several gods had only the withered skeleton of a hog. Though the smell was to me very disagreeable, and the sight of the idols laughable, I did not let the islanders perceive it, that I might not offend them; but I was the more astonished when the priests themselves made me observe the caricatures, felt their noses and eyes, tried to imitate, in various ways, the distorted faces, and laughed heartily at their wit. Near a hut stood two complete statues, whose sexes could be distinguished, clumsily as they were carved; between them a pole was fixed in the ground, the point of which had been hung with bananas. The woman, turning her face to the man, seized with her left hand the fruit, while he stretched out his right hand towards it. On seeing this, every one must think of Adam and Eve; and I was very sorry to have nobody with me to explain this allegory. The priests made me notice that both statues, which had their mouths wide open, were furnished with a row of human teeth. One of the small chapels was covered round about with mats; from this proceeded the noise of the muffled drum, sometimes interrupted by the lamentable cries of a man; and the whole made so unpleasant an impression on me, that I was glad to go away. On my return, I found in front of a house, a large assemblage of ladies, who had placed themselves round a fire on which a dog was roasting. They politely invited me to partake of the feast, but my time would not permit me to-day.

The female sex, to whom hogs'-flesh is forbidden, eat dogs' instead, which are, for this reason, fed only with fruits. It is the peculiar feature of these dogs, which belong to the species of our badger, that they never attach themselves to man, and are therefore reckoned among hogs.

The 12th of October. My intention of leaving Woahoo to-morrow was frustrated by Kareimoku, who begged me to stay here some days. He had the next day a taboo, which did not end till the day following, and therefore would not be able to accompany me; besides this, he represented to me that my voyage would be unfortunate if I began it before the taboo. As he had always treated me with so much friendship, I could not refuse this request. The ship was prepared to get under sail; all the provisions were taken in; and, when at last there were a number of animals on board, such as hogs, dogs, pigeons, cats, &c., the Rurick perfectly resembled Noah's Ark.

The 14th of October. We were ready at sunrise to leave the harbour. The captains of the American ships, whose names I here mention with gratitude, William Davis, John Ebbets, Thomas Brown, and Thomas Meek, had sent their long-boats to tow me out. Kareimoku soon appeared: he had just come out of the morai. He called to me *Aroha!* and said, that the god, at his incessant prayers, had promised to protect us on our voyage, that we might reach our own country with quite whole heads, and healthy feet, and he did not doubt a moment that we should have a very fortunate voyage. He brought with him water-melons and fish from his reservoir, and treated us with remarkably more attention than he did the captains of the merchantmen, to whom he behaved with pride. On parting, I presented him with the portrait of Tamaahmaah, which appeared to give him uncommon pleasure. He left us, after cordially pressing our hands, and recommending us to his gods. The young Kareimoku, who had hitherto remained with us, and who had been entrusted with the present for the king, received one of my embroidered uniforms, and he rejoiced aloud when he put it on. Soon after our friends had left us, a fresh land-wind arose, we spread all our sails, and steered S.W. by W. As I resolved to look for the Cornwallis islands again on this voyage to Radack, I directed my course thither.

The 20th. In the morning our ship was visited by a very large number of snipes; according to our reckoning, we must soon reach Cornwallis' Island, and found at noon, after a good observation, the latitude 16° 45′ 12″ N.; longitude, ac cording to the chronometers, 169° 16′ 37″. The accuracy of our longitude is guaranteed by the observations made, for several successive days, between the sun and moon. I steered to west, a few miles more north than last year, because I supposed that the islands could not lie so far to the south as laid down on Arrowsmith's chart. At two o'clock in the afternoon, after we had been saluted by a number of sea-birds, land was announced, which appeared at a distance of thirteen miles in W. by N. ½ W. Only a small round hill was visible to us. An hour later, at a distance of eight miles, we already saw that this hill formed the northern part of a low island, the whole circumference of which might be about a mile. A mile north of this island a second was seen, just as low, and still smaller. While we were taken up in the survey, the sailor on duty exclaimed, "There are rocks under the ship!" I immediately had the ship turned to south, and we happily escaped the danger of wrecking on these rocks, which we, dazzled by the sun, had not previously observed. The distance from land was here five miles, the rocks were scarcely two fathoms under water, and close to them the lead could not reach the bottom. To judge from this, the shoal was either of very small extent, or it was the point of a coral-bank with which this island seems to be surrounded far into the sea, as is proved by the breakers which we afterwards discovered to the N. and E. I steered, after having escaped this danger, towards the islands, to try to approach them from another side; but this was also in vain, for we were soon convinced, by the colour of the water, of the impossibility of touching at them. Seamen may therefore take warning not to come too close to these islands, which, as I plainly saw through the telescope, consist only of naked rocks. On the hill which was first visible, I observed a white spot. Towards evening, we left this abode of the birds, and steered to the southward. We found the latitude of the hill 16° 45′ 36″; longitude, according to the chronometers, 169° 39′ 21″ west. Declination of the magnetic needle, 9° 47′ east.

The 21st of October. We concluded we were near land, as a great number of ducks from the N.W. flew to the S.E., where we lost sight of them. According to Anson's chart we were in the neighbourhood of Basso de la Villa Lobos. On Arrowsmith's new chart of the south sea, no shoal is marked, but we have reason to believe that there is one here.

The 30th. I had directed my course to Otdia, and this morning, at eight o'clock, we got sight of the island of Ormed, belonging to this group. Kadu's joy at the sight of the well-known land, was indescribable, and he could not conceive how we had found these islands again, after wandering about for so long a time. The wind, which during our whole voyage had been E., and E.N.E., changed, to our great surprise, to S.E., as this phenomenon is quite unusual in the tropics; black clouds arose, which I, however, did not regard, as they moved but slowly, and I therefore continued our course close to the wind, to double the island of Otdia, and to penetrate to-day through

Schischmareff Strait, to our friends in the group. But fate would have it otherwise. When we were within only five miles of Otdia, which lay to the west of us, the clouds covered the whole sky, the rain poured down in torrents, and a violent wind obliged us to take in the top-sails.

Our situation was critical, because the wind, which was too violent to reach the high sea by tacking, drove us towards the land. I still hoped that it would decrease, and we should be able to withdraw from the cliffs, when suddenly a second gust arose with the fury of a hurricane, by which the mast would, without doubt, have gone over board, if we had not quickly taken in all the sails. These gusts, accompanied with rain, lasted for above an hour. We saw the dangerous cliffs near us, and when we could already calculate the moment of our destruction, the fury of the wind abated. We did not neglect instantly to set all the sails, to stand off from the shore. While our situation was the most dangerous, a whale sported round our ship, which seemed to wait for our near destruction with impatience. It was one of those rapacious animals, with their enormous jaws and teeth, which the Aleutians call *Plawun.* How this fish, which is generally found in the north, had come so near the equator, I cannot conceive.

The bad weather had no influence on the barometer. The wind, after a few hours, again veered to the east, but blew violently during the night, and we tacked in sight of land.

The 31st of October. At day-break we took our course to Schischmareff Strait, which we reached at ten o'clock, followed by some gusts of wind. In a boat under sail, which we soon overtook, we recognized our old friend Lagediack, who, as soon as he saw us, made the most comical gestulations in the joy of his heart, always crying, *Aidara, Totabu, Tamisso, Timaro!* As we were under full sail, he could not come on board; he, therefore, contented himself with proceeding to Otdia, where he begged us to follow him. Kadu had resolved not to show himself to his half-countrymen in the canoes, but to surprise them by his presence on shore; but his excessive joy frustrated all his plans. The Radackers were scarcely near enough to speak to him, when, to their great astonishment, he sprung up, exclaiming, "Look here! I am Kadu! do you know me still?" They then began a lively conversation, in which he probably told them the most wonderful adventures, for their long-drawn *O—h* was frequently repeated.

At five o'clock in the afternoon, we cast anchor in the same place where we had been before. Lagediack immediately came, loaded with cocoa-nuts, accompanied by some savages, who were strangers to us. As soon as he came on board, he gave himself up entirely to the joy of seeing us again; he danced and sung, ran up to us, embraced us all by turns, and at last took a wreath of sweet-scented flowers, which he had just twined, from his head to put it on mine, continually exclaiming, *Aidara!* His comrades imitated him in every thing, though we were strangers to them. After the intoxication of his joy was in some measure dispelled, Lagediack came up to Kadu, who was a very remarkable person in the eyes of them all. They formed a circle round him, in the middle of which he was obliged to sit, and immediately words flowed from his lips, his eyes sparkled, and the faces of the audience strongly expressed the sensations which his long narrative had excited. We were at length obliged to interrupt the stream of his eloquence, which had already made him foam at the mouth, as we wished to know what had occurred during our absence from Radack. I was astonished that Rarick did not visit us, and on my inquiry after him, learnt the following particulars. A few days after we had left the group of Aur, the old chief Lebeuliet came to Otdia, as he supposed that we had left much iron there; he forced the inhabitants to give up a part, and sailed back to Aur, having previously taken the three goats which were still living, from the Goat Island. Some months after, Lamary, who had proceeded by way of Ligiep and Ailu, came here with his fleet from Udirick, and to this king of the Radack chain, the people were obliged to give their last iron, and all the things received from us. I asked why they had not opposed the unjust command; but they replied, "Then Lamary would immediately have killed us all." The king had remained here two months, to prepare Mogan for the army which was to attack Mediuro. When he sailed away, he provided himself with bread-fruit and cocoa-nuts, of which he left only just enough for the inhabitants barely to support themselves. Rarick, Langin, Labugar, and several of our acquaintance, had gone with him, and nobody remained on the island except women, children, and some old men, the number of whom was certainly not more than fifteen. The army was assembled, but it had not yet come to battle, as Lamary wished to wait for the attack of the enemy; but he intended, if they did not soon come, to attack them himself.

I inquired after our garden, and learnt that the rats had destroyed every thing, except a few roots, which had succeeded very well, till the great rat, as he called Lamary, had come, and, notwithstanding all their intreaties, had carried off every thing. Though I could not but regret the destruction of our garden, I still hoped that our seeds on the other islands had succeeded better, and showed them the different animals and plants, which I had designed for them. They were all much pleased, and, particularly, Lagediack, who could not refrain from repeatedly embracing me. With Kadu's assistance,

I hoped to instruct them in rearing and attending the plants which were now to be put into the ground. Our orange-trees, which we had brought in pots, were in the best condition; the vines, as well as the potatoes, taro and yam-roots, had already young leaves. M. Von Chamisso, in his observations on Radack, will give a list of the plants. The hogs which we had left here had perished; they had probably died of thirst.

The 1st of November. As our time would not permit us to make a long stay here, the roots and plants were brought on shore to-day, the old garden was put in order again, and M. Von Chamisso took the trouble of planting almost every thing with his own hands. The islanders were obliged to be all assembled, to receive here their first instructions, and Kadu was now, as interpreter, an important personage. But he scarcely gave himself time to perform his duty faithfully. He always thought of his adventures, which he was obliged to relate, thus drawing off their attention from the garden to himself, so that I was at length forced to order the orator to be silent. To make the Radackers acquainted with the taste of the different roots, I had brought with me some of every species boiled; they found them all very agreeable, particularly the potatoe, of which every one wished to have some. I distributed a large quantity of them, which they gratefully received, and I was delighted at the thought that these good-natured people, who were in want of provisions, would, perhaps, in future, thank me for their prosperity. When these islands once produce in superabundance taro, yams, and potatoes for their inhabitants, the cruel custom of murdering the children will cease, and they will at least be less frequently engaged in wars, as they are at present caused only by the want of provisions. Some water-melons, which I had brought with me from the Sandwich Islands, they found very excellent, and desired some of the seeds. I willingly distributed them, but cautioned them against the rats, and Lagediack resolved immediately to make a garden on pillars, to protect it against them. In the afternoon five goats and three cats were put on shore, which I committed to Lagediack's care. The islanders ran together to admire them, particularly the cats, and their astonishment was without bounds, when they on their first step on shore caught some rats, which ignorant of the danger, ran into their mouths. I presented Lagediack with two hens and one cock.

As I intended to leave Otdia in two days, I remained on shore during the night with Chamisso and Kadu, to enjoy the society of our friends. After we had put the garden in order, we sat down before Lagediack's house on the grass, surrounded by the islanders, who tried to amuse us with songs and drums. During our absence they had composed songs of praise on us, which they now sung to us. The song on *Totabu* made the beginning, then followed *Timaro*, *Tamisso*, and others; though I did not understand the sense of their poetry, they were agreeable to me, as they descend as traditions from father to son, and, as such, may be repeated to future navigators by a younger generation. The supper was brought on shore, and we took it in the presence of our friends, who regarded us with looks of interest. Kadu, who supped with us, explained to them the use of the several utensils, and must have expressed himself very wittily, as they laughed heartily. He had, during his stay of nine months with us, improved so much, that he could not but feel his superiority; nevertheless he liked to be with his old friends; kindly instructed them, made presents to their children, and showed this attachment in every way. Much as he prided himself in his European dress, he immediately laid it aside here, and particularly banished his boots and shoes, which he disliked very much. He had very soon distributed all his treasures. During supper Lagediack sat next to me, and ate with an excellent appetite. A dish went round the circle of the spectators, and every one began to take out a dainty with his long nails. The company thought the boiled yams and potatoes very excellent. Kadu exhorted them, on this occasion, carefully to preserve the roots brought by us, that they might have some in future, and laughed very much when one of the savages showed him a boiled yam, saying, he would not eat it, but plant it to-morrow. He thought the Radackers were still too stupid! The pork likewise pleased them, but they did not like the wine; a glass, which went round, they only touched with their lips. Kadu called them fools, who did not know what was good; they should follow his example as he was a man of experience; at the same time he emptied the glass in one draught. After supper they again danced and played on the drums, and when Kadu stepped into the circle to dance in the European fashion, he excited a general laugh, and Lagediack thought that our dances looked as if we had lost our wits. Before we retired to rest I asked Lagediack whether he was acquainted with the Ralick chain, as he had never spoken to me about it; he said he had often been there, and I again observed on this occasion how difficult it is to draw information like this from savages, if one is not quite master of their language. They will never relate any thing, but answer only to such questions as are put to them, supposing that we are wise, and raised far above them without being instructed in any thing. M. Chamisso also often found great difficulty in procuring information from Kadu. Lagediack now told me if I sailed from Eregup to the S. W. I should in a few days come up to the group of Odja *, which is

* The names of Otdia in the Radack chain and Odja in the Ralick chain are so similar to each other, that care must be taken not to confound them.

said to exceed all the others in extent, as well as population. There is a tradition that a long time ago a ship was at Odja, and left them much iron.

I passed a very restless night. As I was not able to cover myself on account of the heat, the rats and lizards played their gambols on my body.

The 2d of November. The old chief of the island of Ormed visited us to-day, rejoiced like a child at seeing us again, and made me the bitterest reproaches that I had not anchored at his island, as he was now chief of the whole group; the good old man, as generous as ever, notwithstanding the scarcity caused by Lamary, brought me cocoa-nuts and bread-fruit. Kadu had formerly resided a long time at Ormed, and was paternally treated by the chief, and both now showed a really affecting joy at meeting again. He accompanied, with Chamisso, his foster-father, to Ormed, where they intended to plant different things, and would not return till to-morrow. In the afternoon I nailed on a cocoa-tree, opposite toLagediack's house, a copper-plate, on which the date of the year, and the name of the ship were inscribed. Lagediack was much delighted at this memento, which he promised faithfully to preserve, but could not conceive how I could sail away with the Rurick when its name was nailed up to a tree.

On the 3d of November, in the morning, M. Chamisso returned with Kadu, and I was disagreeably surprised with the news that the latter intended to stay here. It was but yesterday that he promised never to leave me, and this sudden alteration of his resolution was quite an enigma, which Chamisso soon solved. Kadu had learnt on shore that his little child in Aur lamented very much after him, ran about in the woods all day to seek him, and could not sleep in the night. This news had softened his paternal heart, and brought him to the determination of remaining here. He seemed still to struggle with himself, when he related it to me with much emotion; but when I, though with a heavy heart, as I really loved him, approved his plan, he resolved to execute it, and promised to rear our plantations with affection, and to call the different plants by our names. Future navigators will therefore find, instead of yams, taro, and potatoes, timaros, tamissos, and totabus. Every one on board the ship would know from his own mouth whether he really intended to leave us; and he told to each, individually, how his child called Kadu in the woods, and could not sleep in the night. The separation was very painful to me, and I could only console myself with the idea that he might be useful here, and would not perhaps long survive in our cold climate. As he intended to leave the ship to-day, because we sailed tomorrow, we all collected presents for him. He looked at his treasures with mute astonishment, and was only afraid that the Radackers could not resist the temptation of robbing him. I did not doubt that Lamary, as soon as he heard of it, would not fail to take from him the greatest part, and to avoid this, left some very considerable presents for him also. The old chief of Ormed and Lagediack were not forgotten. Some hogs and dogs, which I entrusted to Kadu's care, were then put into the boat, and I accompanied him with Chamisso on shore, he having previously taken an affectionate farewell on board the ship. Lagediack received us on shore, gazed with astonishment at the treasures, which were spread out, and was enraptured at the presents given to him. Kadu's riches I had brought into Rarick's habitation, where he concealed them, and the islanders, who were delighted at the sight of them, were, perhaps, already forming plans in secret, for appropriating them to themselves. To protect Kadu as much as possible against such an event, I intended to make an exhortation to all the savages. Lagediack immediately dispatched two criers, who went about the island, and made known his orders, that they should assemble. Some drums were beat, and all the inhabitants of Otdia, men, women, and children, were soon assembled. They were informed that Kadu intended to remain here, and that I was going to speak to them on the subject. The people, full of expectation, formed a circle, in the middle of which stood Chamisso and myself. Kadu, in the mean time, dressed himself in Rarick's house, probably to make a strong impression on the savages, on this solemn occasion. After we had been waiting for some time, he at last came out of the house, with measured paces; he had put on a white shirt, a sabre buckled round his waist which he held naked in his right hand, and his head covered with a straw hat. The Radackers were astonished when they saw him enter with a serious countenance with his murderous weapon, and he sat down with much gravity on the branch of a tree. The sun had already set when Kadu made the following speech, in which he had been tutored. I must previously observe that Kadu, from our accounts, had formed a very high idea of the tamon of Russia, of whom he told the Radackers a great deal. "The great tamon of all tamons," said he, "of the land of Russia, has commanded that Kadu shall remain here, to take care of the plants and animals left here by the Russians. Nobody dare hinder him on pain of death; on the contrary, every inhabitant shall assist him to cultivate the land, for which he is to be rewarded;" though the promised rewards were to arise from their labour itself. I also permitted myself the following fiction, in order to give more weight to the speech: "A large ship will come from Russia in ten months, to bring the Radackers iron, and other necessary articles; but if it finds that the plantations are destroyed, the guilty persons will be punished with death! Let nobody venture to rob Kadu, or to do him any injury;

this crime will also be punished with death." At the conclusion, I promised large rewards to such as should, on the arrival of the ship from Russia, come on board with their new-cultivated fruits. Kadu delivered his speech with much dignity; the islanders promised faithfully to fulfil our wish, and, to make them acquainted with my great power, I had given orders on board, to fire, on a signal being given, two guns, and to throw up a rocket. It was now quite dark; I told the islanders to look at the ship in order to see the fire with which we would punish their disobedience. The signal was given, the cannons thundered, and the poor savages were petrefied with terror; but the rocket caused still more alarm, which, hissing through the air, illuminated the whole island. Lagediack threw both his arms round me, and begged me to put an end to the terrific scene; but Kadu was much pleased at the impression the fire had made, for he now thought himself secure against any attacks. Some presents which I distributed restored tranquillity. We gave Kadu two copper medals with the portrait of the emperor; the one he was to wear himself, and present the other to Lamary in my name. He resolved to bury some of his treasures, and go with the rest to Ormed to his old benefactor. On taking leave, Kadu seemed to be fully sensible how hard it was to part from us, for he cried like a child, and implored me to come again. The attachment of this good-hearted man moved me very much, but I was still more deeply affected with the universal lamentations of the savages at our departure. Lagediack kept close to me, and frequently asked me whether we would really return. Men, women, and children accompanied us to the boat; Kadu went on before with a drawn sword; and the torches, with which they lighted the way, gave the whole procession a very solemn appearance. After we had put off they all sat on the shore, and joined in a song, in which our names were frequently repeated.

The 4th of November. The anchors were weighed at day-break, and we left the group of Otdia with the consciousness of having done good there. We saw through our telescopes Kadu sitting before Rarick's house, with several others looking after us. I recognised him by the white shirt; he waved a white handkerchief as long as I could see him through the telescope.

The wind was so faint that it was not till nine o'clock that we reached Schischmareff Strait; we then went, after having passed Eregup and Otdia, to N.W. by W. ½ W., to discover the group of Ligiep, which, according to the account of the Radackers, must lie in this neighbourhood. The wind was so faint during the whole day that we could still see Otdia at sun-set: we had rain and squalls of wind during the night.

The 5th of November. In the morning, at seven

o'clock, the sentinel on the mast-head called Land! It was the group of Ligiep, which showed itself to us in N.W. by N. in a little low island, which we could approach but slowly for want of wind. At noon, the N.E. point of the island of Ligiep, which forms the northern part of the group of the same name, lay to the N.W. 68°, at a distance of three miles and a half from us. We now had a perfect calm, the ship no longer obeyed the helm, and the current which sets here strongly to the west, drove us to the land. The Rurick was scarcely a mile from the breakers, and we were just about to put the boats into the water to rescue her from the approaching danger, when a faint north wind helped us out of the difficulty. We now overlooked the whole group, which, though considerably smaller than the others, perfectly resembled them in other respects. Its greatest extent from N.E. 45° to S.W. 77° was fourteen miles and a half. A boat under sail, with ten men, came up directly to us from Ligiep, through a passage formed by the reefs. As the wind did not assist them, they had recourse to their oars, soon overtook us, but ventured only within thirty fathoms to look at us. When we addressed them in their language they were greatly astonished, talked much among themselves, came quickly up to us, and asked, where we came from? "From Otdia," said I: they repeated with surprise, "From Otdia! from Otdia!" and, at length, asked, whether the Tamon Totabu was on board. When I answered their question in the affirmative, and presented myself as such, all fear vanished; they fastened their canoes to the Rurick, and hastily climbed on deck. Lamary, who had lately visited this group, had told them of the Rurick, and must have given them a favourable account of us, as they approached us with such assurance; and this unsuspicious confidence gave us much pleasure. The natives of Ligiep are tall, robust, and handsome men, by which they are advantageously distinguished from the other Radackers. Kadu had previously told me that they lived principally on fish; and this may, perhaps, be the reason of their stronger make. Our guests had, as it appeared, put on their finest clothes. Their dresses were entirely new, their hair smeared over with cocoa-oil, and tastefully tied up, adorned with shell-wreaths and feathers; in their ears they wore large rolls of tortoise-shell, an ornament which I did not see frequently in Otdia. Upon the whole, they seemed better off, and more lively than the natives of that group. Their first care on coming on board was to make us presents. A Tamon, very much tattooed, laid cocoa-nuts at my feet, and put his shell-wreath on my head; the others did the same with our gentlemen, and we soon had not guests, but intimate friends, on board, who were quite at home. They ran about with looks of curiosity, admired every thing, and likewise felt themselves most attracted by the iron:

they inquired after Kadu, and whether we had brought him back. They now received presents in their turn, were much astonished at our liberality, and tried to express their gratitude by begging us to visit them on shore, where the most beautiful *Ridginis* (women) should receive us. The islanders showed us a passage on the western part of the group, deep and broad enough for our ship; but as I did not intend to visit Ligiep, I would rather have learned something respecting the Ralick chain. I asked a Tamon where it lay, and he pointed to the west: I asked farther, where the island of Kwadelen lay; he pointed again to the west; and I now confidently hoped not to miss the island-group of Kwadelen, as last year the chief of Ailu had pointed out the same direction. After the islanders had been for an hour with us, a high wind arose; we were obliged to part from them, and I took my course to the west, along the group, to finish the survey of it. When the sun was setting we had already sailed round the western part of the group, and steered to the west in the hope of discovering the Ralick chain. We found the latitude of the middle of the Ligiep group to be 9° 51′ 30″ N.; longitude, according to the chronometers, 190° 46′ 30″ W. Declination of the magnetic needle 10° 56′ E. I called the group after our worthy commodore, Count Hayden. During the whole night we continued our course to the west. It rained, and violent squalls frequently obliged us to take in the sails.

The 6th of November. At day-break we all eagerly looked about, confidently hoping to discover the Ralick chain, but in vain. At mid-day we observed the latitude to be 9° 42′ 56″; longitude, according to the chronometers, 191° 52′ 40″; we had, therefore, advanced from Ligiep one degree of longitude to the west, and I was afraid I had already crossed the chain, as such low land is easily overlooked. When the sun set, and no land yet appeared, I was obliged, though with a heavy heart, to abandon the farther search for the Ralick chain, which would have cost me too much time. The northern monsoon in the China sea was requisite for me to reach Manilla, where the ship was to be repaired; this same monsoon was also to carry us through the Straits of Sunda. As in this hitherto-unvisited part there might be other islands besides the Ralick chain, I ordered the ship to lay to during the night; and, on the 7th, at day-break, continued the course to west, but this day was also spent in the vain expectation of seeing land. The current had carried the ship, in twenty-four hours, eighteen miles to the west.

The 9th, we observed the latitude 9° 32′ 54″; longitude, according to the chronometers, 197° 22′ 24″, and sailed over the place where the Island of Casbobus and the thirty-six islands of the Spaniards are said to lie, without discovering even the least indication of land being near. I still continued my course to the west, either to come up to the island of Hogelon, or perhaps to make some new discovery, as this part has not, to my knowledge, been examined by any navigator. The water had already, for some days past, become of a blueish colour, and has since become a hundredth part salter than usual. During the night we had violent rain, gusts of wind, and lightning.

The 11th: latitude 9° 19′ 56″; longitude, according to the chronometers, 201° 25′. We were now at the place on which Hogelon's Island was said to lie, but looked about for it in vain, and I believe I may affirm that it does not exist. As well on the 11th, as on the 12th, a number of distances were taken between the sun and moon; the longitude calculated from which, exactly agreed with the longitude by the chronometers.

The 13th, at noon, we were in latitude 8° 59′; longitude, according to the chronometers, 204° 24′. The water is still uncommonly blue. I suppose that the sea, from the Ralick chain to this place, and perhaps still farther to the west, is of less considerable depth than east of Radack. The Six-thermometer, which I let down to-day, during the calm, gave the temperature of the water in the depth of the sea much colder than it is on the other side of Radack, and in all other parts under the tropics which lie in this latitude.

OBSERVATIONS.

Temperature of the air - - -	85° 0′
Temperature on the surface of the water -	87 0
At a depth of 100 fathoms - - -	56 2

Transparency of the water thirteen fathoms.

We were troubled with a calm for several days, I therefore steered more to the north, to reach the region of the monsoon, and take the direct course to the island of Guahon, one of the Ladrones.

The 14th: latitude, at noon, 9° 21′; longitude, according to the chronometers, 204° 44′. The calm permitted us to make the following observations with the Six-thermometer.

Temperature of the air - - -	84° 00
Temperature on the surface of the water	83 00
Temperature at a depth of twenty-five fathoms - - - - - - -	77 00

The 15th. There is still a calm. The latitude at noon, 9° 25′ 48″; longitude, according to the chronometers, 205° 00′ 45″. To-day, a large shark was caught, and in his belly was found, to the great astonishment of the crew, a seal-skin cap, lined with cloth, which one of our sailors, a few days before, had been obliged to throw overboard, to avoid the jokes of his comrades. This cap, which had nearly the fate of Casem's slipper, was scarcely to be recognized, on account of the oil and blubber, and had become very heavy. Notwithstanding this, the owner declared he would not part with it in spite of all their jokes; and he in fact wore it, till, one day, in taking in a sail on the top-yard, it fell from his head, and luckily, only on the back of another; had it fallen on his head from this height, he might have been killed. He now threw it overboard the second time, and thought that the shark must have been the devil himself, who had tried to catch his poor soul, by bringing back the unfortunate cap.

The following observations were made to-day, with the Six-thermometer.

Temperature of the air - - -	85°	70′
Temperature on the surface of the water	87	40
At a depth of 15 fathoms - - -	79	00
Do. - - 50 - - - -	59	00
Do. - - 69 - - - -	51	40
Do. - - 101 - - - -	49	50

I observed to-day the transparency of the water with a white plate, and found that it was visible at a depth of twenty-seven fathoms: the previous observations of this kind had been made with a piece of red cloth. I daily let down the Six-thermometer about eighty fathoms, to observe the difference, when the water should have resumed its dark azure colour.

The 20th. When we were in latitude 10° 42′; longitude, according to the chronometers, 209° 51′,

I remarked a sudden change in the colour of the water, which was now again azure. The Six-thermometer gave us the temperature, at a great depth, as much warmer, and confirmed me in my opinion, that we had had a less deep sea since leaving Radack than is usual between the tropics. It may be possible, that the bottom of the sea from the Philippine islands to Radack forms a ledge. The Six-thermometer gave us, to-day, at

a depth of eighty-six fathoms - + 63° 00′

The 15th, it gave us, at sixty-nine fathoms - - - + 51 40

The temperature of the ocean, at a depth of eigthy-six fathoms, is, therefore, here, ten degrees warmer than it was on the 15th, at sixty-nine fathoms, though, according to the general rules, it ought to be colder; and this circumstance clearly proves, that the ocean must be here much deeper than in that part where we were on the 15th. After an examination by Dr. Eschscholtz, the water to-day was found to be a hundredth part less salt.

The 23d. At half-past eight o'clock in the morning, we saw from deck, the island of Rota, or Zarpane, in the north-west 68°, but only its southern part, as the northern part consists of very low land. Even the southern part was, however, scarcely visible, though we were distant from it only nineteen miles. I steered up to Zarpane, as I wished to determine its longitude. At a quarter before eleven, we descried the island of Guahon, twelve miles distant, to the W S. W. As, according to Arrowsmith's chart, we could not expect to see Guahon before passing the longitude of Zarpane, this latter is, probably, laid down incorrectly. Zarpane may, perhaps, be a little higher than Guahon. I estimate the first at from three to four hundred toises high. On the western part of its southern point lies a small island, of inconsiderable height. We found the latitude of the southern point of Zarpane, to be 14° 00′ 58″; longitude, according to the chronometers, 214° 39′ 46″ I steered directly between the two islands, keeping in the middle of the strait, which I took to be free from danger, but which is not at all the case, as I afterwards learnt. In the middle of it, a little nearer to Zarpane, there is a shoal, which lies only three fathoms under water, and might have been dangerous even to our Rurick, which draws only twelve feet water, if the waves had run high; large ships have therefore to be cautious, even when the water is smooth, of striking on this bank, as it is not laid down on any chart. The Spaniards in Guahon affirm that, in a high sea, the breakers may be seen over this shoal. At noon, Zarpane lay to the north-west 20°, distant from us eight miles and three quarters, but did not afford an inviting prospect, as we saw only naked rocks. I now steered up to the northern point of the island of Guahon, which we reached at four o'clock. Here the country seemed to be more picturesque, and we regretted that we should not be able to visit it to-day. There is no chart of Guahon to be depended on: the town of Agaña was only known to me by description; and as it was too late to look for it to-day, we kept out at sea during the night.

The 24th. As soon as the morning dawned, I again directed my course to the northern part of the island, along the west coast of which I intended to proceed to the south, till we discovered the town of Agaña. The northern part of the island of Guahon rises perpendicularly from the sea, to a moderate height, and runs to the south in a straight line, as far as the eye can reach: a beautiful forest of variegated green covers the upper part of the island, and affords a very pleasing view to the mariner. The wind blew so high that we were obliged to take in the top-gallant sails. At eleven o'clock we had the northern point of Guahon behind us, and were under the lee of the island where the monsoon, checked by the height of the land, died away to almost a perfect calm. A breeze now and then, which put the ship a little in motion, wafted to us the most aromatic odours from the shore, to which we were near. The seaman who has long been deprived of land, knows, particularly, how to prize these pleasing sensations.

At the distance of half a mile to the east of us lay a promontory, (called by the natives Tuloberspit,) which, on sailing from the north point to the south, is the first thing you meet with, as from here to that place the land runs almost in a straight line. The promontory may be easily recognized, by its uttermost point being formed by a cylindrical rock, which rises perpendicularly from the sea. From this point the land forms to the south a deep bay, divides itself into several small creeks; and the scenery here becomes delightfully picturesque. Could I have transported myself back to the time when Magellan discovered these islands, the Rurick would long since have been surrounded by many canoes, with happy islanders. This was not the case now; the introduction of the Christian religion has not diffused here its benign blessings; for, since that time, the whole race of the natives of the Ladrones has been extirpated. We looked in vain for a canoe, or a man on the shore; and it almost seemed as if we were off an uninhabited island. The sight of this lovely country deeply affected me. Formerly these fertile vallies were the abode of a nation, who passed their days in tranquil happiness; now only the beautiful palm-groves remained to overshadow their graves: a death-like silence every where prevailed.

I intended to cast anchor a quarter of a mile from Tuloberspit, but was prevented by the great depth and the coral bottom. To our great joy, we now saw a man on the shore, who appeared to us naked and of a black colour; but he ran hastily into the woods when he saw our ship. Soon after, we discovered a large boat, which I recognized to be an European by its mode of rowing, and soon came up to it. A young Englishman, of the name of Robert Wilson, who held the place of pilot in Agaña, had been sent by the Governor to conduct us to a safe harbour, if we intended to stay at Guahon While he steered S. W. along the coast, we could, without interruption, continue the survey. We soon saw the town of Agaña, in the south, which is pleasantly situated on the shore; and, on the west, leans against a tolerably high hill, which forms a cape (here called the Devil's Cape); on the summit is a fortress, in which, while still at a distance, you see a small white house. I expressed to Wilson my wish rather to anchor before the town of Agaña than in Umatack Bay, where, as is well known, Malespina, some years ago, had so very bad a birth, and was answered, that it was only in former times, when the present harbour was unknown, that ships anchored at Umatack or Agaña, where the situation was in several respects so bad, that many ships have perished there. At twelve o'clock, when the town already lay S. E. of us, we saw a canoe under sail coming up to us, in which Wilson, through a telescope, recognized a deputy of the Governor. I immediately lay-to, and we soon had the pleasure of seeing on board, Don Ignatio Martinez, lieutenant of artillery, who, as he did not know our flag, asked to what nation we belonged. He was much astonished to see Russians before him, but still more so when he heard that we were making a voyage of discovery, and showed even double the politeness natural to the Spaniards. After he had written down the name of the ship, and mine, he took his leave, and hastened to communicate this important news to the Governor. The boat in which the officer visited us was very similar to the canoes of Radack. Its structure, the sail, the way of managing it, and even the inhabitants of this place, naked and of a dark colour, transported us for a moment to Radack. We learnt from Wilson that there are here several such canoes, which they obtain by barter from the Carolinians, as they sail quick and close to the wind. The chain of islands, known to us by the name of the Carolinas, lies at a considerable distance south of Guahon; and, for some time past, there comes yearly from there, notwithstanding the distance, a small fleet, which gives the Spaniards shells, corals, and trifles, in exchange for iron. Kadu told me frequently of Tautua, chief of Ulle, who sailed to Wagal to barter there for *Lulu* (so the Carolinians call iron). We could now no longer doubt the truth of Kadu's assertion; the island of Wagal, so often mentioned by him as lying to the north of Ulle, was no other than Guahon, where the chief Tautua was very well remembered. We now hastened to get into the harbour: a long narrow tongue of land, called Orot, on the western point of Guahon, before which lies a small rocky island, forms the entrance. The whole harbour, as may be seen from our chart, is formed of coral reefs, in the same manner as Hana-rura, in the island of Woahoo. The harbour is protected, towards the north, by a small low island, called Appapa; it is overgrown with thick bushes, and appears, on sailing from the north, to be connected with the continent. From this island there runs, to the west, a coral reef, the uttermost point of which, with the cape of St. Carlos de Orote, forms the entrance of the harbour, which is a mile and a quarter broad; but in the middle of it there is a shoal *, which is, however, not dangerous for small ships; but I would advise large ones not to keep in the middle of the passage, but rather to the south of the bank, and as near as possible to St. Carlos de Orote, where the water is deep enough for the largest vessels. At two o'clock in the afternoon, we entered the harbour by the northern passage; its depth was five fathoms and a half, the bottom consisted of coral, and we were in a basin of water as smooth as a mirror, where ships can also lie at anchor; but as the depth is considerable, and the bottom not good,

* A Spanish galleon, from Acapulco, richly laden, and bound for Manilla, which called here to take in provisions, is said to have been wrecked on this bank a short time ago, and the whole cargo went to the bottom.

they generally go to the inner harbour, which is one of the safest in the world. An east wind obliged us to tack to the entrance of the inner harbour, an attempt which Wilson thought dangerous, on account of the many coral banks, and which no ship had hitherto ventured. He advised me rather to cast anchor here till the wind had veered to the west, which it usually does in the morning; but as this would have taken me too much time, and as we had already acquainted ourselves with the nature of coral banks at Radack, we resolved rather to attempt to work through directly. According to the old custom, a sailor was obliged to be on the mast-head, another on the bowsprit, and the mate in the shrouds. We were timely warned of every danger, and tacking, to the great terror of Wilson, who had already protested against all responsibility, fortunately entered in safety the inner harbour. As the entrance is so very narrow the Rurick was obliged to be warped in: all hands were put in motion; and, at five o'clock, we were inside of the harbour La Caldera de Apra. We here found a deputy from the Governor of Agaña, who, in a very polite letter, invited me and all my gentlemen into the town, and had already sent us some mules, which waited for us on the opposite shore of the island Appapa, near the village of Piti. I accepted the invitation with pleasure; left Lieutenant Schischmareff to anchor near the fortress of

St. Cruz, which lies on a small island in the harbour, and went on shore, accompanied by our scientific gentlemen, and Mr. Wilson. We had to row one mile and three quarters to the village of Piti, as we were obliged to make many turnings on account of the shoals; saw a small two-masted vessel, belonging to the Governor, lying at anchor; and, besides this, no other ship in the harbour. Wilson, the mate of the two-masted vessel, assured me, that years sometimes elapsed, in which no ship came here. The sun was near setting when we landed at Piti, from whence we went to the adjoining village of Massu, where the mules were in waiting for us: the Governor had sent his horse, the only one in the island, for me. As we were still three miles and a half from Agaña there was no time to be lost; I mounted my horse, the other gentlemen got on their mules, and we rode off in the highest spirits. The scenery was very romantic, and seemed a paradise to us, after so long a voyage; and at the same time, the air, with its odoriferous perfumes, had such a beneficial influence on us, that we all felt ourselves strengthened.* Anson represents the island of Tinian, which he reached in a sickly state, after a long and fatiguing voyage, as one of the most romantic in the world. Had he landed here, nobody could have

* The Spaniards resident here universally affirm, that the air of Guahon is very healthy; and that people attain an extraordinary old age.

denied this assertion, as has been done by several mariners with respect to Tinian.

The village of Massu consists of about fifteen houses, which are built in a straight line, and the spaces between them filled up with gardens. Their structure is different from any we saw on our voyage. The house, which is from eight to ten feet square, rests on four pillars, raised five feet above the ground; the floors and walls are made of bamboo canes, which are placed so far apart from each other, that you can put your hand between them, which gives the whole house the appearance of a cage, where you may see every thing passing in the inside without entering. This construction is very well adapted to the climate: the wind passes through the house, and cools and purifies the air; the roof, thatched with rushes, protects it against the rain; and the pillars against vermin; but the appearance is extremely ludicrous, particularly if the family is in it. The half-naked inhabitants of Massu kindly saluted us in the Spanish language. A large stone cross before their village, and a small one, which they wear round their neck, showed the Christian faith. The inhabitants of Guahon are called by the Spaniards *los Indios*. They are all Christians; are partly descended from natives; but most of them from people brought hither by the Spaniards from Mexico and the Philippines, after the original race was extirpated.

The way we rode was narrow, but very beautiful; the mountains lay to our right, varied with picturesque scenery. We rode through palm bushes, and sometimes through a wild but romantic country, in which we admired the variety and luxuriance of the vegetable kingdom. When the sun set, the moon guided us on our way; in this light, the unknown trees and bushes made a singular impression on us, and we often fancied that we saw a monster at a distance, which, when we approached it, changed into a sago palm. The air had become cooler, I therefore put spurs to my horse, the gentlemen followed on their mules, and at eight o'clock in the evening, after passing through two other villages, we arrived at Agaña, at Wilson's house, where we dressed ourselves, to pay our respects to the Governor, Don Joseph Medinilla y Pineda, captain-general of the Marianes or Ladrones, who received us in full uniform, and with the greatest politeness. After I had acquainted him with the object of my voyage, and had told him that I landed here with the hope of obtaining fresh provisions, he promised, with the greatest readiness, to provide me with every thing the season of the year produced, and regretted that it was not now the time of fruit, which was there in abundance. He immediately gave me a proof of his desire to serve me, by ordering his aide-de-camp to send to the Rurick, at break of day, fresh meat, fruit, and vegetables, and daily to supply the crew with them. The governor is, in fact, the only real Spaniard; the other officers, and even the priests, are natives either of Manilla or Mexico, and descendants from Spaniards. He is about forty years old, and, notwithstanding his bad health, an agreeable companion, and a particularly polite host. He must know how to make himself useful to the government, as his administration had been prolonged for three years longer, though, according to the law, the Governor of a Spanish possession, can only fill his office for three years. I had little difficulty in conversing with him by Wilson's assistance, but I tried in vain to turn the conversation on the Marianes, respecting which, I wished to learn many things; but, mysterious as all the Spanish governors are in this quarter of the globe, he always contrived to give the conversation another turn. But he took the more care to please our palate. After he had several times offered us tea and chocolate, he conducted us to a table, richly laid out with fruits, confectionary, and the choicest wines. We enjoyed ourselves very well, as we thought this was to be our supper. Scarcely had an hour elapsed, when we were shown into the dining-room, where the table, covered with the richest dishes, awaited us. We did not know whether it was the custom here to eat without intermission, or whether they gave only the Russian stomach credit for such powers of digestion, but we soon observed that they all ate with an excellent appetite. At table I met the vice-governor, as he is called here, Don Louis de Torres; and this amiable man particularly interested us, as he had visited the Carolinas, and even the group of Ulle. He related to us a great deal about them, and promised to give us, in writing, the observations he had made there.*

De Torres was here when the Carolinians, in 1788, visited the island of Guahon, in a great number of small boats. The savages pleased him very much by their mildness; he received them with kindness, and persuaded also the governor, who dismissed them loaded with presents, and since this time they have had the courage to come every year. They told Torres that they had previously had a commercial intercourse with the inhabitants of this island, and only given it up on hearing of the settlement of the white men, and having themselves been witnesses of their cruelty. In 1788, after a long time had elapsed, they undertook this expedition to barter for iron. Torres asked them how they had found their way here, as the distance from Ulle to Guahon is above three hundred miles; they answered, that the description

* As M. Von Chamisso is master of the Spanish language, he undertook, with pleasure, to copy the remarks on the customs of the Carolinians, and to lay them before the public, in the third volume of my voyage. They are so far highly interesting, as scarcely any thing is hitherto known of the Carolinas. For this reason, I resolved to remain here longer than I first intended.

of the way was preserved by them in their songs, and after this their pilots had found it. It is really very remarkable that they did not miss an inconsiderable island like Guahon, when the stars and the songs were their only guides, in a voyage of three hundred miles. When the Carolinians visited Guahon in 1788, they promised to return in the following year. They kept their word; but on their voyage back were overtaken by a furious storm, and found a watery grave, so that not one of the brave seamen saved their lives, and, after this circumstance, de Torres waited fifteen years in vain for his friends, to whom he had become attached on account of their gentle dispositions. In the year 1804 the American ship Maria, from Boston, took in provisions at Guahon: the captain of it, Samuel William Boll, undertook, with the super-cargo, Thomas Borman, a voyage to the Carolinas, where he intended to make the attempt to catch *biches de mer* *, and de Torres embraced this opportunity of visiting his friends, as the captain promised to bring him back to Guahon. The Maria sailed in July, and the first

* *Biches de mer* is a species of large snail without shells, and is found in warm climates, particularly near coral reefs. This snail is much sought for in the Chinese market; the Chinese consider it as a delicacy, and pay a great price for it, as they ascribe to it the power to restore health and vigour. I have eaten these snails at the Governor's, but found them quite insipid. In Radack they are found in abundance, in the basin of the island-group; but the natives do not eat them.

group at which she touched was Ulle. Torres found here several of his old acquaintances, who piloted the ship into the group; and this was the ship of which Kadu spoke to us: the name Borman, of which he made *Marmol*, and *Louis*, as they called de Torres, occurred in one of his songs, which the Carolinians had composed, that these men might not be forgotten. The preservation of remarkable events in songs is therefore common to the Carolinians and Radackers; but I do not know whether the latter praise their heroes as is the case with the Carolinians. Torres inquired why his old friends no longer visited him in Guahon. They then told him of the fleet which had gone there fifteen years ago, and, as it had not returned, they concluded that their countrymen had all been murdered. Torres of course declared that no harm had been done to their brethren in Guahon, but that a raging storm had overtaken them the day after their departure, and probably destroyed the whole fleet. The Carolinians regretted the unfortunate accident, but were rejoiced to hear that no murder had been perpetrated, as they had imagined; they promised to visit Guahon in the following year, and kept their word. Since this time, eighteen canoes assemble every year at the island-group of Lamureck, thence they sail to Fojo (a desert island, according to the description, lying to the north of Lamureck,) which they reach in two days, rest there, and then the fleet sails in three days to Guahon. The whole voyage is therefore concluded in five days. They visit Guahon in April, and commence their voyage home in May, or at the latest in June, as, after that, the S. W. monsoon becomes dangerous. Their boats are of such a kind, that they upset with the least carelessness, which happens several times a-day in such a voyage; but as they are expert swimmers and divers, it has no other consequence than that they have a hearty laugh. On such an occasion they all immediately leap into the water, right their canoe, and swim along with it till they have baled out the water with their hands. It is much worse when the outrigger breaks, for then they are not able to preserve the equilibrium of their canoe on account of their narrow construction. There, however, does not pass a voyage but this accident happens, and they repair it while swimming, which requires several hours. A European would hardly be able to hold out in a voyage of five days, constantly washed by the waves. The Carolinians are often even fourteen days in this situation, without any food except a few cocoa-nuts, as their canoes cannot carry a cargo; and sea water for their drink, as much as they please. They have, when there is a whole fleet, generally two pilots, who are only of the lower class, but far exceed the higher in sagacity, and are often elevated to the rank of nobility for their merit.

Some years ago, a fleet, which was only a day's voyage from Guahon, was overtaken by a violent storm, and driven far out of its course. When the storm abated, the two pilots disputed; the one affirmed that Guahon still lay to the west, and the other thought the contrary, as the S. E. storm had driven them so far, that the island must lie to the E. of them. Both had always possessed the confidence of the islanders; they were now at a loss whose advice to pursue, and, in the end, the fleet divided into two parties. The party sailing to the W., probably found its grave in the waves, as nothing was afterwards heard of them: the other, after it had laboured several days to gain the wind to the E., happily reached the island, and the pilot, as a reward for his services, was made a tamon.

When the Spaniards took possession of the Mariane islands, most of their inhabitants fled to the Carolinas. Louis de Torres saw, on his voyage, many islands belonging to the Carolinas, and drew a chart of the whole chain, which M. Von Chamisso likewise copied, and will add to our atlas. The present Governor has given himself much trouble to gain the confidence of the Carolinians, and has proposed to them to settle at Guahon.

As the Governor had not room for all of us in his house, he only kept me and Chamisso; the other gentlemen were kindly received by the authorities in the town. The Governor's house has two stories, and is quite adapted to this climate. The interior was high and spacious; the wall towards the north, furnished with sliding-frames, but not closed, except when the sun shines in. Instead of glass, the shells of the pearl-oyster are fixed in frames, which certainly admit light, but not the hot rays of the sun: the wall to the south has no windows. Care had been taken that our night's lodging should be convenient, but the constant quarrels between the dogs and cats, would have disturbed our sleep, as the latter frequently sought refuge in our beds, had we not been extremely fatigued by our ride. A kind of small green lizard, which runs whistling up the walls during the night, and sometimes takes the liberty of crawling on the bed, and creeping under the quilt, is found here in all the houses. Dogs, as well as cats, are in great numbers both in town and country, and care is taken of the breed of these animals, as the rats do great mischief every where. The dogs are used in hunting the stag, which abounds here; it is a small species brought by the Spaniards from the Philippines.

The 25th of Nov. We were scarcely awake when the Governor sent to invite us to take chocolate. After we had partaken of it, I expressed a wish to see the town. I was not allowed to do so till I had made another breakfast, which perfectly resembled a dinner.

The town of Agaña, which should properly be called a village, is situated on a beautiful plain, some hundred paces from the shore; to the right and left are fine palm groves: in the south, a high mountain forms the back-ground; from its summit, hang large spreading trees, which overshade a part of the town, and give it a picturesque appearance. An inconsiderable stream, which flows through the town, supplies the inhabitants with water. The houses, which are constructed in the same manner as in the villages, form several regular streets. Only seven or eight houses are built of coral stone, which belong either to the government, as that of the Governor, or to the civil officers.

On the eastern part of the town is a considerably large church, and a convent; the whole clergy consists of two priests, who are natives of Manilla, and descendants from the Malays. It is said that, always after a certain time, generally every twenty years, a violent storm rises in the S. W., which causes the sea to run so high that the town is overflowed, and the inhabitants obliged to fly to the mountains. Only the stone houses are able to resist the fury of the water; the bamboo cages are all destroyed. Two fortresses of coral stone defend the town; one of them lies on the shore, but has not yet any guns; the other lies to the west, behind the town, on an eminence, has a few guns, and seems to be built chiefly to restore tranquillity in case of a riot; but as they have no powder, as I was informed by the Governor, I do not see the use of the two fortresses. The town has two hundred houses, and contains 1500 inhabitants, who, as I have already said, derive their origin from Mexico and the Philippines. There is only a man and his wife on the whole island, of the original branch. With the death of these two people the race of the old Ladrones will be entirely extinguished. The Governor had the kindness to show them to me, and our painter took their portraits. The military consists of militia, and appears to be in good condition: the officers are natives. The soldiers, who are obliged to provide their own clothing, looked very neat, though they are bound to give a part of their scanty allowance to the priests. When a native intends to marry, he must previously give the priest a Spanish dollar, who pays no regard to the prevailing scarcity of money. On my walk, the Governor showed me several canoes, which he had purchased from the Carolinians; and told me, on this occasion, of the great dexterity of these people in swimming and diving. When the galleon, already mentioned, perished, some Carolinians, who were here at the time, fetched the barrels filled with piasters out of the cabin of the ship, which was lying several fathoms under water; it was affirmed that they remained there half an hour.

The whole island-chain of the Marianes, except Guahon, is uninhabited. The North Americans, who carry on the fur trade, between the N.W.

coast of America and Canton, had chosen the islands of Agrigan and Saypan as a place of rest on this voyage. To be sure of finding there, in future, fresh provisions, they brought several families from the Sandwich islands, who were obliged to cultivate the ground, and tend the cattle; and they would have succeeded in their future voyages, in being supplied with fresh provisions from these two islands, without payment; but the Spaniards had scarcely been informed of it, when soldiers were sent there, who made the poor islanders prisoners, and destroyed their plantations. I have seen these islanders at the Governor's; they seemed contented with their fate, and were much rejoiced to hear something about their native country. The Governor had learnt that the Americans had founded a new colony on Agrigan. It is a question how long it will last.

When I took leave of the Governor, in the afternoon, I was obliged to accede to his kind invitation, to visit him the following day. M. Chamisso remained on shore, and I returned to the Rurick, accompanied by Dr. Eschscholtz. We were delighted at the romantic scenery, and were shaded, by overspreading trees, from the scorching rays of the sun. We halted in the villages, and the people were always ready to refresh us with juice drawn from cocoa blossoms, of a very pleasant taste. The sportive, cheerful disposition of the South Sea islanders is no longer found among these people; they have been too long oppressed, and all their actions are indicative only of subjection. They are entirely under the command of the Governor; and though the government demands no taxes, their condition is but poor. The present Governor is a good man, who treats the poor half-savage Christians like his children; the former one was a tyrant, whom they never approached without trembling; and, perhaps, on this account, they are still distrustful.

Tobacco is universally prized; men, women, and children, constantly smoke segars; they have also betel in their mouths, which gives a disagreeable red dye to the teeth and lips. In all the Spanish settlements the government has the sole right of planting tobacco; but every body is allowed to do it in Guahon.

After a ride of two hours, we arrived at Massu, where the boat was in waiting for us. The natives had observed that some of our sailors did not wear a cross round the neck; and they thought that they must be bad Christians. At five o'clock we reached the ship, which was now lying in the inner harbour, near the fortress of St. Cruz. Lieutenant Schischmareff had already begun to lay in a stock of water in a convenient place, which I have marked on my chart. The following particulars are to be observed: the boat must be sent off at high water, that it may reach, without difficulty, the mouth of the river; here the barrels are to be immediately unloaded, floated into the river, but not filled till the ebb has carried the salt water out of the river; in the returning tide the boats must be loaded, that you may go back when the water is at the highest. The coast, near which the water is taken in, has a singular appearance; the bushes dip their points into the sea, from which new roots shoot up, which entwine themselves in the higher bushes; and this produces a thick green wall, that covers the whole shore. The tree as easily takes root in water as in the ground, and it looks as if the vegetation sprung from the sea.

The 26th. While I was on shore, Captain Taitano, commandant of the fortress of Orota*, had visited the ship, and invited Lieutenant Schischmareff to come to his house. It was situated behind the Cape of Orota, in the village of Agat; and went to visit him, accompanied by Wilson, and likewise to take a view of the town. The whole crew wished to go on shore to-day, and received permission, on this occasion, to gather as many oranges in the groves as they could carry. We

* The fortress of Orota lies on an eminence of Cape Orota, and defends the entrance of the outer harbour; its situation is so very advantageous, that it can neither be cannonaded, nor fire a shot in vain. A mile to the east of this, a second (St. Louis) lies on the shore, which defended the entrance of the inner harbour, but is now decayed through negligence. The fortress of St. Cruz, on which I made all my astronomical observations, is very small, built of coral stones, and contains only three six-pounders, and is not even guarded.

landed on the southern part of the harbour. To reach it, it is necessary to be acquainted with the channel, as it is rendered dangerous by the many shoals; but, as Wilson was our pilot, we had no accident. A narrow foot-path led us across the tongue of land through thick bushes, and we soon saw the sea, and were in a large open bay *, in which lie the three small islands laid down on my chart. We went then through an avenue of palm-trees into the adjoining village of Agat, two miles behind which we saw a round mountain, the highest in the whole island. Taitano received us very politely, and the romantic scenery with which his house is surrounded, made an agreeable impression on us, and we all returned to the ship, where the crew had already arrived, as highly delighted with their walk as with their oranges. The sailors saw in the forest a stag, and several lizards five feet long. Besides cats and dogs, stags are the only quadruped we have seen. There are no poisonous reptiles, or snakes.

The 7th of November. I left the ship in the afternoon, with Schischmareff, to visit the Governor, as I had promised; at Massu we found a horse and a mule, and were received on our arrival with as much politeness as the first time. There were a number of natives who were to lead up a dance; but as there are no longer any national

* I took the plan of this bay with my pocket-sextant and a small compass.

dances, which we wished to see, the scene was introduced where king Montezuma of Mexico receives Cortez, and amuses him by the dance of his subjects.

The 28th. We returned early to the ship, as I proposed to leave Guahon the following day. Louis de Torres accompanied us, with all the officers, and the Governor, who intended to give me some dispatches to Manilla, promised to follow, and to remain on board during the night. We spent a pleasant evening in the society of the Spanish officers, who all remained on board during the night; the governor had been detained, and did not come till the morning of the 29th. The ship was abundantly supplied with fresh provisions, among which there was even a live ox. We parted with grateful hearts: when the Governor got into his boat, we fired a salute of five guns, and the crew gave three cheers. At eight o'clock we were already out of the harbour.

We found the latitude of the Harbour Calderona de Apra to be - - - - - -	13° 26′ 41″
Longitude - - - - -	215 9 54
Declination of the magnetic needle	5 34 00 E.
According to Marion's observation the harbour lies in longitude - - - - - -	216 10 00
According to Malespina, Umatak Bay lies - - - - -	215 26 00

As the difference in the longitude between Umatak Bay and the harbour of La Calderona de Apra can only be very inconsiderable, my longitude agrees pretty exactly with Malespina.

CHAP. XIV.

FROM GUAHON TO ST. HELENA.

The 1st of December, latitude 16° 31′, longitude 219° 6′. A number of sea-birds indicated the vicinity of an uninhabited island. According to Arrowsmith's chart, there lies a sand-bank in this latitude, which is said to have been also seen by Spanish seamen.

The 6th: latitude 20° longitude 232° 12′. A small hawk was caught to-day. We discovered some days ago a considerable leak in the ship; probably a piece of the sheathing had come off, and the worms, which are so numerous between the coral groups, had pierced through the wood; we therefore wished to reach Manilla as soon as possible.

On the 9th, at ten o'clock in the morning, the island of Baton, one of the Bashee islands, was discovered in the west, from the mast-head, at a distance of twenty-seven miles. I now directed my course in such a manner, that I might be able to come in the Chinese Sea, between the island of Sabtang and the three Balington's rocks. We were favoured by a fresh wind, and, at half past three o'clock in the afternoon, the largest and most easterly of the Balington's rocks lay within seven miles of us, due south. My chronometers gave for its longitude 237° 13′ 30″, latitude 19° 58′ 5″. Upon this I steered more southward, to double Cape Bajador; and, by degrees, the high rocky Babuyan islands came in sight, and we had, therefore, now left the great ocean, after sailing about it for above two years. I was seized with a certain emotion on quitting the South Sea, where we had spent so many unhappy, and yet so many joyful hours; and I now looked upon my voyage as almost concluded. How great is the difference in the atmosphere in sailing from the ocean into the Chinese Sea. Instead of the almost constantly serene sky, we saw here stormy clouds tossed about by the wind, and the horizon enveloped in an eternal veil. The currents had carried us to-day eighteen miles to the S. E., 18°.

The 10th, at noon, we were in latitude 9° 12′, longitude 239° 43′, and we had been carried by the current, since yesterday, twenty-six miles and a quarter to the N. E., 3°. We were obliged to take in all the reefs, on account of the violent wind. As it blew from E. by N. from the land, we sailed briskly up to Cape Bolinao, while we sometimes discovered, through the fog, the summits of the mountains of Luconia.

The 11th, the wind was less stormy, and we found the current thirty-four miles and a half to the N. E., 14°.

The 12th, in the evening, we saw Cape Bolinao, and doubled it during the night. We always kept within sight of land. On the 14th, at noon, doubled Cape Capones, and were now obliged to gain Manilla Bay by tacking, as in this season of the year the wind always blows from the bay. The water in the hold increased very much.

The 15th, at noon, we were near the island of Coregidor, and saw there a telegraph in great activity, which announced our arrival at Manilla. At sunset we had reached the southern entrance into Manilla Bay, and tacked just between the island of Cavalo and the continent, when we were stopped by a large boat with twenty oars, here called Pango. A Spanish officer came on board, and asked, with much politeness, to what nation we belonged, and with what intention we came to Manilla, which he must report to the Governor. There are several similar guard-boats near the island of Coregidor, but they are properly there to prevent the Moors from coming into the bay, who frequently come from the southern islands of the Philippines, plunder, murder, and even make prisoners, whom they afterwards sell as slaves. The officer sent us a pilot to conduct the ship into Manilla; he had, however, little skill in his department. The chart which I had of Manilla Bay was very incorrect. I was therefore obliged to follow the information I had collected from different voyages respecting this bay.

St. Nicholas's shoal is the most dangerous place, but may be easily avoided by care. We tacked during the whole night, but without gaining much, as the wind was calm. In tacking, care must be taken to go so far to the north, that the northern entrance may be kept in sight, as you are then in the current, which in this season of the year always flows out violently. This is the reason why ships can only penetrate through the southern entrance during the N. E. monsoon, and, in sailing out, use the northern.

The 16th. The faint wind still detained us at the island of Coregidor, on which we observed a crater; there probably was formerly a volcano near this place, which, by its fall, formed several small islands and a basin. As the wind began to blow a little we hoped soon to reach the town of Manilla; but the calm which immediately set in, obliged us to cast anchor eight miles from the town, that we might not be carried back by the current.

The 17th. The calm continued. At one o'clock at noon a sixteen-oared boat came up to us. Two officers, deputies from the Governor, who was already informed of our arrival, welcomed us in his name, assuring us how rejoiced he was to see the Russian flag flying in his harbour, which had never been seen there before. The gentlemen said to us many flattering things about our nation, calling it, among other things, the deliverer of Europe. I embraced the opportunity of going to the city in the *pango* with the two Spanish officers, with the intention of paying my respects to the Governor, and asking his permission to sail to Cavite to have the Rurick repaired there. M. Von Chamisso, who is master of the Spanish language, accompanied me, and at four o'clock we reached the road, which is very secure in this season of the year, and where we saw eight merchantmen, under the American and English flags, lying at anchor.

Manilla is situated on a plain, and does not afford, from the sea-side, a very pleasing prospect, as you see only one stone-wall planted with cannon, above which appear the roofs of the houses, and some church-steeples. To be able to land, you must row into the river, which is likewise deep enough for large ships, and at the mouth of which is a shoal, above which there are only fourteen feet water at the new and full moon. The river divides into several branches, and two of them, the banks of which are strewn with villages, lead to the celebrated lake of Bahia, which lies twenty miles farther up the country. The shoal had a very animated appearance; a number of fishing-boats, manned with Chinese and Malays, carried on their profession here; some of the boats attracted our attention more particularly, as they threw a large net into the water by a simple machine managed by two men, which was drawn out in a minute filled with small fish. When we had penetrated into the river, we saw, on its right bank, the town surrounded by a well-built wall; to the left we saw a large village inhabited by Malays, which, as in Guahon, consisted of nothing but bird-cages. After we had landed in a pleasant valley, we traversed a number of dirty streets in the town, between high houses built of stone, where the air is confined and oppressive, till we arrived at the house of the governor, Don Fernando Mariana Fulgeras, who received us with much politeness. He permitted me to sail to Cavite, and the captain of this port was to receive orders to-day to give me all possible assistance in repairing the Rurick. I took my leave to go there as soon as possible, because the north-east monsoon allowed us at the most a delay of two months, which might easily pass away before the ship could be ready to sail. The Governor, who seemed to be an intelligent and agreeable man, begged me frequently to visit him; he offered to assist M. Chamisso if he intended to make an excursion into the interior of the country, and an elegant carriage with four horses brought us to the *pango*, in which we reached the Rurick at seven o'clock in the evening. A light wind had just arisen. I ordered the anchor to be weighed, and we tacked during the whole night to reach Cavite, which lies twenty-one miles south of Manilla.

The 18th. There was so little wind, that we had not advanced much, and it was noon before we arrived at Cavite, where two merchantmen lay at anchor. M. Von Chamisso immediately went on shore to announce our arrival to the captain of the port, who, early on the 19th, sent us two long-boats with kedge-anchors and cables, which warped the Rurick into the arsenal where she was to be repaired. I now visited the captain, Mr. Tobias *, who directly ordered the ship to be unrigged and unloaded; all the things were removed into an empty galleon lying close to her, where our sailors were likewise conveniently quartered, and a house was prepared for us at Cavite, according to the orders of the Governor. Mr. Tobias, with whom I easily conversed, as he spoke French fluently, accompanied me, with a shipwright, to the Rurick, and all the preparations for the repair of the ship were made.

On the 20th we took possession of our house, which was situated on the bank of the river, and pleased us very much. The houses are all furnished, like that of the Governor at Guahon, with a balcony to the north, which can be closed by sliders with mother-of pearl panes.

By the zeal of Mr. Tobias, who put a hundred men to the work, it proceeded very rapidly, though there was a great deal to do, as the sails, tackling, boats, masts, pumps, and even water-butts had be-

* Son of the Governor of the Marianes, mentioned by La Peyrouse.

come quite useless on the long voyage. A small ship, like our Rurick, has the disadvantage that it is not able to carry with it all the things required, and is therefore often obliged to pay a high price for the assistance of others. On careening the ship, it was found that most of the copper-sheathing was unserviceable, and the bottom in many places gnawed by worms. I therefore wrote to the Governor at Manilla, and he ordered Mr. Tobias to have the ship new-covered with copper. We are indebted to the great activity of Mr. Tobias, that, notwithstanding the extent of the repairs to be done, our ship was finished in proper time. I, in the meantime, employed myself in trying my chronometers, and making fair copies of the charts we had drawn.

Cavite, inhabited only by military and Malays, who are used for work, is a fortress, which is not a pleasant place of residence. You must take a long walk before you reach a village, the houses of which are built of two stories, partly in the Chinese, and partly in the Malayan fashion. I went here every day when it became cooler, to amuse myself in the illuminated market, which is always kept after sunset. Hundreds of women, sitting in long rows on the ground, sell all sorts of vegetables, fruits, and the like; and the workmen in the fortress, and even the military, come here to take their supper. The crowd is very great, and as the natives are very musical, scarcely ever laying their guitars aside, after supper, they play, dance, and sing in the open air. Three miles from here, lies the very romantic village of Terra Alta, to which Mr. Tobias frequently drove me in his chaise. The road lies through avenues of mango-trees, resembling our limes, but bearing the finest fruit: the neat and pretty habitations of the Malays, with their plantations, agreeably entertain the traveller. In Terra Alta the scenery is very beautiful, and vegetation luxuriant. Many rich Spaniards have here country seats, as they consider the air particularly healthy.

The 24th. Christmas-eve put all Cavite in motion; priests with the images of saints passed through the streets, the Malays followed the procession, and children ran on behind with lanterns, which were of the figures of various animals. A pleasing music was heard at intervals, but was soon drowned by the noise of fireworks and rockets. In this night nobody sleeps at Cavite; at twelve o'clock all the bells are rung, and the people flock into the church to prayer.

The 25th. To-day we made an excursion, in a boat belonging to Mr. Tobias, to Manilla, where we were entertained by the aide-de-camp of the Governor, as there is not a single tavern here. The Governor immediately sent us two carriages to visit the well-known picturesque environs of Manilla, where we saw a number of richly-dressed ladies and gentlemen on the favourite promenade, both in carriages and on foot.

On the 26th. I was just prepared to pay my respects to the Governor, when he came to us himself, and invited us to dinner. In the morning we visited the beautiful suburb, which is mostly inhabited by rich Chinese, who have their shops there, and know very well how to cheat the Christians.

After dinner, to which all the first people of the town were invited, we rode home. It is not till evening that the higher class of the inhabitants begin to stir; till that time they are occupied in sleeping, eating, and smoking tobacco, which is no where more general than on the island of Luconia; for children, when they are yet unable to walk, already smoke their segars. The women carry their love for it to a still greater extent than the men; they are not content with the usual small segars, but have others made for them, which are a foot long, and proportionably thick. An idea may be formed of a mouth, whose lips can take in such a roll of tobacco. These large segars are called here the women's segars, and it affords the most ludicrous sight, when these elegant ladies take their evening walk, with these burning brands in their mouths. The chewing of the betel is one of the luxuries of the fair sex, and is particularly hurtful, as it is wrapped in a leaf strewed with unslaked lime, and so chewed. The Government, which alone has the right of cultivating tobacco, sells it at 4—5 reals per lb, and the king has a revenue from it, from the island of Luconia alone, of about 300,000 piasters; the rum, which is extracted from cocoa-blossoms, belongs likewise to the king, and yearly brings him in 120,000 piasters.

In the evening we drank chocolate at the Governor's, and had an enjoyment of which we had been deprived of many years, for his two amiable daughters enraptured us by their singing and playing. He persuaded me to return to-morrow to Cavite, by land, over Terra Alta, for which he offered me his carriage. I joyfully accepted this proposal, and we were already, at seven o'clock in the morning, on the pleasant road, which leads through avenues of bamboo and cultivated fields. I, for the first time, saw here such high bamboo canes; but the people in this country know how to make a very good use of them, for they make out of them bridges, houses, and utensils of all kinds. Half way to Terra Alta, lies a convent, at which we stopped, as Chamisso wished to speak with a monk, who had written a history of the Philippines. We remained there to dinner, and rode further, having previously been offered, by a Frenchman of the name of Chaparé, in the Spanish service, his country house, if we would come here more frequently. We continued our journey, and safely arrived at our fortress in the evening.

The 28th. Immediately after my arrival I took care to have my six Aleutians vaccinated; the district-physician received orders to come on board with two children who had the cow-pox, and Dr. Eschscholtz inoculated the Aleutians. On the island of Luconia, the surgeons have the strictest order to inoculate the little children in the villages every week.

To-day Mr. Tobias sent out from the arsenal a small galley-fleet against the Moors, which is done twice a-year. The fleet consisted of nine gun-boats; five of them carried each a four-and-twenty-pounder, and the four smaller ones had ten-pounders: they were all well manned, and amply provided with small arms. The fleet sails to the Straits of St. Bernardino, where it separates; the one-half takes its station in the strait, and the other goes to the northern part of the island of Magindanao. Since the Spaniards have adopted the course of chastising the Moors in their own country, they do not venture so often to penetrate to Manilla.

We had here, with the north-east monsoon, during the day, 23°, and in the night 18° of heat; and while we could scarcely endure this heat, the inhabitants wrapped themselves up in the night in their warm bed-clothes, and called this the winter month. To judge from this, the heat during the southern monsoon must be intolerable; many people then die suddenly when they are heated and expose themselves to the north wind, which sometimes blows in the summer. There is a sickness called St. Lazaro, the most frightful that I ever heard of. They have first a leprosy over the whole body; the limbs rot off, and the unfortunate person, who has inevitable death before his eyes, retains his senses to the last moment in the most excruciating torments. This sickness prevails principally among the poorest natives, and is probably caused by bad fare and uncleanliness. An hospital in Manilla, which is supported by government and the rich people of the town, is filled with people ill of this disease, and a monk, who has had the superintendance of it for twenty years, was of opinion that there was no remedy against this punishment of God, as he termed it. I made the observation, that the poor creatures had dirty woollen clothes next to the skin, and received spoiled meat for their food; but when I asked whether clean linen and healthy fare would not ameliorate their condition, I was answered, that it would cost too much!

Lazy as the natives are in their work, they possess the more ingenuity in cheating, particularly strangers. Their only pleasure, of which they are passionately fond, is cock-fighting; and they bring up cocks for this purpose, which they always carry about with them. In every village there is a house built by government, in which all the cocks are permitted to fight, but only on Sundays and holidays; the spectators pay a real for admittance, but the owners of the cocks are obliged to pay four reals, and the king receives the profits. The stage, on which nobody is allowed to come, is surrounded by two rows of boxes. When the contest is to begin, the sum betted for is deposited; each owner now puts his cock, which is furnished at both feet with knives two inches long, upon the stage, and the contest is often decided by the first, but more frequently by the third or fourth, onset. The vanquished cock is most cruelly treated by his master; for, to punish him, he immediately plucks out all his feathers. Great sums are lost, for the spectators sometimes lay wagers; and even if one of them were to lose the last piece of raiment on his body, he quits the theatre as cheerfully as he entered it.

The 12th of January, 1818. I rode to Manilla, in order to inspect the provisions intended for us, and resided at Dr. Amador's, who had been recommended to me by the Governor of the Marianes. The following day I visited the Archbishop of Manilla, Don Juan Antonio de Zulaybar, and gave the old man much pleasure, as he had never seen a Russian, and yet, as he said, esteemed the nation very much. At dinner the Governor told me the following circumstance, which is said often to happen here: sometimes horses run about at liberty in the interior of the country, which are subject to the singular fate, that a bird makes its nest on the upper part of their tail; as soon as this happens the horse grows lean, and does not recover even after the bird has flown away with its young. The Governor seemed to be perfectly convinced of the truth of this phenomenon, which I should have declared to be a fable on hearing it from a less-informed and intelligent man.

I visited to-day, in the suburb, the segar-manufactory, which was formerly a convent for monks, where I found 2000 women and 350 men busily employed.

On the 14th, I rode back to Cavite; the repairs of the ship were almost concluded, and the Rurick was already rigged.

The 26th. I brought my chronometers on board, and made preparations to sail next day to Manilla, where I had to take in biscuit and other provisions. The Governor sent our painter a girl, who was descended from the mountaineers in the interior of the country, that he might take her portrait. These, as it is well known, were formerly the sole inhabitants of the Philippines, and after being driven away by the Malays, they lead a peaceable *Nomade* (life) in the mountains. They do not like to have any thing to do with Christians, and do not wish to be baptized.

The 28th. Yesterday at noon we left Cavite; and a few hours afterwards cast anchor before

Manilla. The Governor visited us to-day; we expressed to him our gratitude, and, on leaving the Rurick, he was saluted by fifteen guns. Captain Guerin, of the French ship Eglantine, wished to follow me to the strait of Sunda, as he had no chronometer on board, without which the Chinese Sea is dangerous. I therefore acquainted him with the necessary signals, and we both left, on the 29th, the beautiful and fertile island of Luconia.

The 3d of February. The fine weather and a fresh N. E. monsoon had hitherto favoured our voyage; at nine o'clock Pulo Sapata lay distant from us fourteen miles due west, and the chronometers gave for the longitude of this island 251° 00′ 6″. In the evening the sky was covered, dark clouds announced a stormy night, and the gusts of wind became so violent, that we were frequently obliged to take in the sails.

The 6th. At four o'clock Pulo Aor appeared in S. W. 25°, twenty-two miles from us. The Eglantine had remained so far behind that we were obliged to wait four hours for her, with the topsails reefed. I now tried to pass Magellan's shoal and Gasper Island to the west, not to come into Gasper Strait, which appeared to me more convenient and safe than passing it to the east, as many mariners do.

The 8th, at six o'clock in the morning, we crossed the line in longitude 253° 9′. We saw to the right, at the edge of the horizon, a ship under sail, which, as it came nearer, I recognized by its sails and structure to be a pirate ship belonging to the Malays.

I soon perceived that the hostile ship sailed better than ours, but always trying to intercept our course, kept constantly at a certain distance, probably not to create suspicion, and not to attack till the darkness of night. I knew that the inhabitants of Banca and Sumatra infest these parts in large vessels, holding three hundred men, and not seldom board merchant vessels, which they then plunder, and murder the crew. Some of our gentlemen thought I carried my apprehensions too far. I was not to be dissuaded, but immediately put the ship in a state of defence; the guns were double-loaded with grape-shot and ball, the matches lighted, and the whole crew stood on the quarter-deck, armed with sabres and fire-arms, ready to defend it to the last drop of their blood. When it became dark, two sailors were placed on the bowsprit, and precisely at eight they cried "A light!" It was exactly in the part where our course was leading us, and seemed to be not far off, but immediately vanished, and I now had some sails taken in, to be able to manage the ship better in case of an attack. We sailed slowly on, and a death-like silence prevailed, which was fearfully and suddenly interrupted by the cry, "A light! a light! the ship is close up to us!" I had myself seen the light, which instantly vanished.

The vessel, notwithstanding the dark, was easily distinguished, and had we kept our course two minutes longer, the Rurick would certainly have been boarded, which was probably the intention of the robbers. Firmly resolved to conquer or die, I immediately ordered the right side of the ship to be turned against the enemy, which was at the most twenty fathoms from us, and the guns were instantly fired; the grape-shot and balls must certainly have taken effect at so small a distance, which must have been very unexpected to the pirates, and done them much damage; for our cannons were scarcely fired, when they took another course, and we still heard, for some time, different voices crying all together. Thus we escaped a danger which, with less precaution, might have cost us our lives, and even this would perhaps not have availed us if the robbers had not been so imprudent as to let us see the light. When Captain Guerin, who had again remained half-a-mile behind, heard our cannonade, he thought that we had fired as a signal of distress, and that we had been wrecked, and he turned his ship that he might escape the shoal. I made a signal that I wished to speak with him; the Rurick lay-to till the Eglantine came up to us, and after I had related to him this adventure, we continued our course.

The 9th. At eleven o'clock A.M. we saw from the mast-head the island of Gasper, in S. by W., and at noon, it was thirty-seven miles from us in

S. W. 8°. Our longitude, according to the chronometers, was 252° 52′ 40″. The wind was so faint, that we approached the island but slowly. We observed a strong current to S.E.; and, notwithstanding the dark, sailed, at eleven o'clock at night, round the western side of the island, at a distance of seven miles; and, at midnight, when it lay about eight miles to the north of us, we cast anchor, as the passage between Pulo Leat and the island of Banca, is dangerons in the night; the Eglantine was likewise obliged to cast anchor. The depth was sixteen fathoms, over a bottom of gray sand: the current ran to the S.E. one mile and a half an hour.

The 10th. At day-break we weighed anchor; the wind blew fresh from N.W., but soon veered to the W.; at noon we had Gasper Strait behind us, and sailed with a faint wind up to the Two Brothers. I did not learn till afterwards, that a year before the English frigate, Alceste, on board which Lord Amherst went as ambassador to China, was wrecked on his voyage back on a hitherto unknown shoal, which is said to lie in the neighbourhood of Pulo Leat. We did not observe it; for as the northern part of Pulo Leat appeared dangerous to me, I remained at a considerable distance; but we sailed so close by the western point, that we could distinguish people on the shore with the naked eye.

The 12th. At six o'clock in the morning, the Knob Mountain, or Montagne D'Apres, in Sumatra, was discovered from the mast-head; at seven o'clock we could clearly see it from the quarter-deck, in S.W. 12°; and three hours after, also the Two Brothers, in S.W. 9°. At noon we had sailed between the Two Brothers and Sumatra, and steered towards the Strait of Sanda, but were obliged to cast anchor on account of the calm. The Two Brothers lay ten miles from us, N.E. 23°.

The 13th, a faint land-wind brought us but little forwards. At noon, the North Island lay three miles from us, N.W. 12°, and when the wind arose, at two o'clock, I took advantage of it to reach the island of Zupften, where I cast anchor two miles from the coast of Sumatra. The North Island lay seven miles from us, N.E. 14°: a mile and a half north of our anchoring-place, lay three little low islands, thickly overgrown with forests, and not laid down on any chart.

Near to us was a boat, in which people were actively employed in fishing, and seemed to pay no attention to us; but I perceived that they observed us very closely, and when they came a little nearer to us, I threw them a knife, which they received with a friendly nod of the head. They gave us to understand, by signs, that they would bring us a large animal from shore, and immediately disappeared behind the three islands. The islanders were lean, and of a dark colour, and their teeth dyed black, gave them an ugly appearance. Their heads were covered with large straw hats, of the Chinese shape, and their dress consisted of old nankeen shirts, from which I concluded that they had much intercourse with Europeans. The boat was the trunk of a tree hollowed out, and provided with an out-rigger. In an hour the islanders returned with an enormous large turtle, which they laid on the back, and tied across the boat; two monkeys and several parrots were sitting on its belly. The one to whom I threw the knife gave us to understand that the turtle, which he called *Curpat*, should be drawn up into the ship with a rope, and two sailors could scarcely drag up this prodigious animal, on the flesh of which the whole crew lived two days. When the turtle was on board, which I received as a return for the knife, the half-naked islander himself came on board the Rurick, with a small bundle in his hand; without saying a word, or making any motion indicating a salute, he sat down on the quarter-deck, and untied his bundle. We all looked with surprise for the treasures he was about to produce, when he pulled out a pair of very old silk small clothes, embroidered with gold, which he put on without any ceremony. After having finished his toilette, he assumed an important air, gave me to understand that he gave me the large turtle, frequently repeating the word, *Present.* I observed that he wished to have presents, and gave him beads, scissors, knives, and some other trifles, which pleased him, but did not wholly satisfy him. His chief desire was to have a pistol, which he named intelligibly, and powder, which he called in his language, *Belbedil;* and as I did not give it him, he seemed to regret that he had been in too great a hurry in parting with the turtle. In the mean time, another canoe arrived with five men, of whom one spoke a little Spanish and English; the cargo likewise consisted of monkeys and turtles, but which they would only part with for piasters, pistols, and belbedil; and when they received a piaster, they sounded it for a long time, to see whether it was really silver. We bought several monkeys, among which there was one that had the face of an old woman, and was of a species which our naturalists declared hitherto unknown in natural history. They also let us have some fowls, and one might certainly be easily supplied with provisions, if one would stay a few days. After the islanders had disposed of almost all their goods, they left the ship, and we regaled ourselves with the prospect of the beautiful country, and the mountain of Sumatra, which rises proudly into the clouds.

The 14th. At day-break, we got under sail, and the current carried us rapidly to the Strait of Sunda. I had determined to take the channel between Zupften islands and the current rocks, which we had already passed at seven o'clock;

but here the wind became faint, the current carried us towards the rocks, which were close by; and our situation would have become hazardous, had not a fresh wind, which suddenly arose, rescued us from the danger. The number of the Zupften islands is laid down on the chart as five, but we counted eight of them. My plan to sail between the islands of Crocotoa and Tamarin, was frustrated by the unfavourable wind, which obliged us to tack. At noon, the pier on the island of Crocotoa lay S.W. 60°, and the pier on the island of Tamarin N.W. 20°. At six o'clock, at noon, we had reached the Strait; I did not wait longer for the Eglantine, which sailed too slow, and had now passed all danger, and therefore continued my voyage without delay.

The 15th. At eight o'clock in the evening, we succeeded in sailing entirely through the strait.

The 16th. The wind became steady from the east, and began to blow high, with which I steered to S.W. and S.W. by W., to leave the Cocoa Island to the south.

The 2d of March we had reached, with a fresh monsoon, the latitude of 22° 2'; and longitude 289° 40'. At six o'clock in the evening the ship was so illuminated by a large globe of fire that we could distinguish every object as in the day; it proceeded from the eastern side of the Pleiades, and descended perpendicularly to the horizon: the whole appearance lasted only three seconds.

The 4th of March. We crossed at midnight the tropic of Capricorn; and the 12th, when we were in latitude 29° 19′, longitude 313° 26′, it seemed that the monsoon which had hitherto favoured us would leave us. The wind turned to the north, and this sudden change was very sensible in the temperature: the wind blew cold upon us, while the air was oppressively hot. Black clouds spread over the horizon in the north, where we observed constant lightning: the barometer fell, and I expected a violent storm, which frequently occurs on the southern point of Madagascar, where we then were. At midnight, while we were just sailing with a fresh wind at seven knots an hour, there came suddenly a calm; the waves, which were running in different directions, agitated the ship: I thought the current must be violent here; and it was proved by observations, the next day at noon, that it had carried us forty-eight miles and a half to the north, 72°. During the calm we saw quite close to us (it was so dark that we could scarcely distinguish objects at fifteen fathoms distant) a large three-masted ship; and as the Rurick could by no means be made to obey the helm, we expected to run foul of the other ship, which would have been our destruction; however, the wind, which had so often delivered us when the danger was at the greatest, rose now at the proper time, and fortunately separated us.

The 17th of March. Latitude 32° 40′; longitude 325° 36′. A violent gust of wind from the west obliged us hastily to take in the sails, not to lose our mast; the rain poured down in torrents, the thunder rolled awfully over our heads; to the right and left the lightning fell around us into the sea, but our Rurick was protected by God. Such squalls are said to be frequent at the Cape of Good Hope; and the mariner has reason to dread them, as with the least negligence the ship may go to the bottom.

The 27th. After we had been detained near the Cape for some days by violent storms from south-west, the wind veered to the east, and we were already to-day off the bank. I had taken my course to its outermost edge, to take advantage of the power of the current, which runs to the west. Our meridian observation gave for the latitude 35° 18′; longitude 337° 4′. We found the current seventy-two miles to the south-west 66°, therefore three miles an hour.

The 29th. At noon, when we were in latitude 34° 55′, longitude 339° 2′, we descried at the same time high land in north-east 32°, and north-east 10°, lying to the east of Cape Lagulas.

The 30th. At noon we doubled the Cape of Good Hope, and steered along the coast to Table Bay. At six o'clock Robben Island was distant from us fourteen miles, lying north-east 73°. There was but a little wind, and we but slowly approached the channel which is formed by Robben

Island and Cape Verde, and leads into Table Bay. The sun had already set, and we had not yet reached the channel: I resolved to pass it, notwithstanding the dark, which was not easy for me, as I had never been here before; we, however, succeeded, and cast anchor in Table Bay at one o'clock in the morning. Misled by several fires on the shore, I had not indeed reached the place where ships usually anchor; but I thought myself very fortunate to have arrived here, as immediately such a violent storm arose that we were obliged to throw out another anchor, and to lower the top-mast, with the yard, to the deck.

The 31st. It was not till day-break that we perceived that we had not anchored before Cape Town, but in the eastern part of the bay, distant three miles from the town, where nineteen ships were lying at anchor. I found it remarkable that we had still to contend against the storm from the south, while the ships at anchor before the town had a calm, and farther up even a breeze from the north, consequently quite in an opposite direction: the difference in the winds is caused by the Table Mountain. A pilot now came on board, we weighed anchor, and we had scarcely gone a mile to the west when the storm ceased, and a northerly wind carried us to the town, where we cast anchor among the other ships. I immediately went on shore to pay my respects to the Governor, Lord

Somerset, who was, however, at his country-seat, and would not return till the next day.

The 1st of April. After I had been honoured by a visit from the captain of the French corvette, L'Uranies, M. L. de Freycenet *, who was about to make a voyage of discovery, I went to Lord Somerset, who invited me to visit him at his seat, Newland, which lies five miles from the town. The Table Mountain was covered with white clouds, a certain sign of an approaching storm, and it began to blow with such violence in the evening, that nobody could come to the ships though they were only fifty fathoms from shore, and I was obliged to remain for the night on shore.

The 2d. It was impossible for me to reach the ship yesterday, but it was still more so to-day; I was even forced to give up my visit to Lord Somerset, as it was generally asserted that the villa could not be reached in this storm, which threw up immense masses of sand, and even little stones.

The 3d. The storm raged with still more fury, no person ventured out into the street. My boat, which was lying on the shore, was driven away by the wind, and very much damaged: several ships in Table Bay were torn from their anchors, though they had lowered the topsails and yard to the deck. This storm reminded me of the hurricane which

* Freycenet is known from Baudin's Voyage, whom he accompanied as lieutenant.

had done me so much injury, and I thought myself very fortunate in having entered the bay in the night. When the storm had at last abated, and I returned to the ship, I found it every where full of sand, and as if crystalized by the salt spray of the water. The barometer never announces a south-east storm here; on the contrary, fair weather may be expected, for before the violent storm began it was 31°, and during the storm rose to 31° 10′. The north-west storm is said, however, to have a great influence on the barometer.

The 4th. To-day we had again the most beautiful weather. I visited the corvette L'Uranie, and M. Freycenet had the politeness to show me all his instruments, and other things worth seeing, among which is a distilling machine, by which sea-water may be converted into water fit for drinking, particularly attracted my attention. This machine, which occupies a space of twenty feet in breadth, and ten in length and height, was in the front part of the lower hold of the ship: in one day it produces fresh water enough to supply one hundred and thirty men for three days, which requires only a very inconsiderable quantity of coals. I had not the pleasure of being introduced to the young Madame Freycenet, who accompanies her husband, and is the first lady who has made a voyage of discovery.

The 5th. To-day, at length, I dined at the seat of Lord Somerset, which is very pretty, and every where bears the stamp of Dutch industry. Here I was introduced to the worthy Colonel Warre, who invited me for to-morrow, to accompany him to Constantia, which I wished to see.

The 6th. I rode with Colonel Warre to Constantia, six miles distant, but which has been too often described for me to say any thing new about it. I can only impart to the reader the hope that the delicious Constantia wine will in future be cheaper, as an Englishman has planted a new vineyard, which, from the nature and situation of the ground, produces a wine equal to any in Constantia. I found the environs of Constantia very beautiful, and the wine, a hundred years old, with which we were treated, very delicious. The director of Constantia assured us, that he was not at the trouble of travelling about, to become acquainted with the different nations of the earth, for all came to him; but a Russian officer he had never yet seen. On my return, I observed a number of small pretty birds, resembling the humming-bird; and Colonel Warre, on this occasion, told me a very remarkable circumstance, which he frequently witnessed, when he made an excursion into the interior of the country, accompanied by some Hottentots. The Hottentots, who have a very quick sight, try to observe a bee flying home with its honey, and pursue it; but they often would not succeed in following the bee, were they not assisted by the birds above mentioned, which perceive the intention of the men. The bird now pursues the bee, and gives the Hottentots, who pursue both, a signal by a whistle, where the honey-comb is; and when they have taken out the honey, they throw some to the bird, which they call the honey-eater, as a reward for his service.

The English have raised a regiment of Hottentots, which is said to be very good, as they are all excellent marksmen, bear great hardships, and can endure hunger for a long time. The Caffres, who often disturb the colony, and are kept in subjection by the English soldiers, fear them very little, for they are accustomed to say, "What kind of fellows are you, you can neither hear, see, nor run! But it is difficult to escape the Hottentot: he sees like a hawk, runs like a hare, and his balls always hit the mark."

The Hottentots, compared with the Caffres, are very short, and, in fact, form a distinct race of men, differing entirely from the other African tribes, but they are very much esteemed on account of their mildness and attachment to truth.

I did not return till the next day to the Rurick, where M. Von Chamisso had already returned from his excursion to the Table Mountain, where he collected a number of plants.

The 13th. When water and fresh provisions had been taken in, we left Cape Town on the 8th of April. The 13th, in latitude 30° 39′, longitude 345° 33′, the calm permitted us to make the following observations with the Six-thermometer.

Temperature of the air - -	68° 0′
Temperature of the water on the surface of the sea - - -	67 0
At a depth of 200 fathoms - -	49 5
- - - 50 - - - -	60 8
- - - 25 - - - -	66 0

The 21st. From the commencement of our voyage to this day, we had, reckoned from the meridian of Greenwich, gone 360 degrees from E. to W., and, consequently, lost one day; I therefore advanced the 21st to the 22d, and Tuesday was made Wednesday.

The 24th. We saw the island of St. Helena at a distance of 50 miles in N.W. I resolved to remain here one day, that the Russian commissary, Count Balleman, might have an opportunity of sending letters by me to his country, and approached towards evening an English brig of war, which cruises here, and strictly examines all ships going to St. Helena. The officer came on board; but before he entered my cabin cocked a pistol, which he had concealed in his sleeve: he advised me to remain during the night near the island, that they might announce our arrival, at day-break, by the telegraphs, and that we could then proceed to James Town.

The 25th. I steered up to the S.E. point of the island, which the English call, on account of its conical mountain, the Sugar-loaf Point. The brig made signals, the telegraph answered, and it did not occur to me that a cannon-ball, which was hissing over our heads, might perhaps be intended for us, as the officer on duty gave me permission to sail to the road; but when, regardless of the Russian flag which we bore, a second ball flew between the masts, I laid the ship to, to expect an explanation, and soon after the Lieutenant of the ship of the line, Conqueror, offered to conduct us himself to the road, and said that the battery had no right to fire on us. We now got again under sail, and in the same moment the third ball passed over our heads; I again lay to, and the officer left us, with the promise that we should receive permission, at eleven o'clock, to go into the road; but when it had not yet arrived at 12 o'clock, I had the flag struck, thanked for the friendly reception by firing a gun, and directed my course to the island of Ascension. The longitude of this island has been laid down so differently, that I wished to go quite close to determine it accurately by my chronometers.

CHAP. XV.

FROM ST. HELENA TO REVEL.

THE 30th, we saw the island of Ascension at fifty miles in N.W. 40°. At noon, it was still twenty-two miles from us: we sailed round its eastern side; and at half-past five in the evening saw its middle due west at the distance of one mile and a half: the chronometers gave its longitude 14° 22′ 30″. We now took our course to the equator, which we crossed on the 6th of May, longitude 20° 26′. The current which had regularly carried us from St. Helena to the S.W. to-day changed its course to the S.E. We bade farewell to the southern hemisphere, and solemnized the day on which we crossed the line for the last time.

The 3d of June. At five o'clock in the morning, we saw Flores, the most westerly of the Azores, sailed past their eastern side, took our course to the English channel, and on the 16th of June cast anchor in the evening, before the town of Portsmouth. Business obliged me to go to London, where I had the honour to be introduced to the Prince Regent, and to our Archduke Nicolai Pawlowitch. I here left with the inventor several boxes of patent meat, as a proof how well it had kept, and to whom all navigators indisputably owe their gratitude.

On the 30th, we left England, remained some days at Copenhagen, and on the 23d of July, with an indiscribable emotion, saw Revel again, which I had quitted three years before, full of pleasing expectation, though not without some fear. Fortune favoured my voyage; and my joy, at the sight of my beloved native city, expressed itself in an ejaculation of gratitude to Heaven.

The 27th of July, I left Revel, and on the 3d of August, 1818, cast anchor in the Newa, opposite to the palace of Count Romanzoff.

ANALYSIS

OF THE

ISLANDS DISCOVERED BY THE RURICK

IN

THE GREAT OCEAN.

BY

CAPTAIN COMMODORE KRUSENSTERN.

ANALYSIS

OF THE

ISLANDS DISCOVERED BY THE RURICK.

SOME accounts of the discoveries of Lieutenant Kotzebue have already appeared in the public prints; they have, however, occasioned remarks, as if most of the islands seen by him had been previously known, and therefore the honour of their discovery did not belong to the Rurick. This essay is therefore intended to contain a short analysis of the islands seen by Lieutenant Kotzebue, which will enable every one to judge how far this assertion is well founded, and whether the voyage of the Rurick has done any service towards the extension of geography.*

As the Great Ocean has, during the last fifty years, been crossed in all possible directions, as well by merchantmen, as by ships expressly sent out to make discoveries, a very scanty harvest can be expected for a mariner who now undertakes a

* The Royal Society at Göttingen has already published, in its learned journal, an extract from an account, which I had the honour to present to it some years ago, on the first discoveries of Lieutenant Kotzebue. In this analysis I shall indeed speak of the later discoveries, but shall be obliged to repeat many things from the account given to the Royal Society.

voyage of discovery; but it might perhaps be as important, and for the geography of the South Sea indisputably more interesting, to solve the doubts which still exist respecting some former discoveries which have never since been recognised, and thus to regulate the geography of the Great Ocean, than to discover here and there a new island. The vanity of the nation to which the navigator belongs might perhaps be more flattered by the discovery of some new islands, than by finding again some more ancient discoveries; the gain for science is however less, especially if the position of the newly-discovered islands is not determined with astronomical precision.

Among the problems still to be solved in the great ocean, the islands discovered by the Dutch in the seventeenth, and in the beginning of the eighteenth century, are those which have chiefly occupied the attention of geographers. Dalrymple and Burney in England, and Fleurieu in France, have expressed very different opinions on their probable existence; particularly Fleurieu, who has written a very elaborate treatise in the third volume of his excellent edition of Marchant's Voyage, on the discoveries of Admiral Roggewein; and many geographers have been induced, by the well-known talents of this learned navigator, to adopt his opinions as the most correct. It has, therefore, long been the wish of geographers, that the regions in which the discoveries of Le Maire, Schouten, and Roggewein lie, should be examined: a task which was not easy, as there is not a more dangerous ocean for navigation than this: it is covered with low coral islands and reefs, scarcely projecting above the surface, and winding in different directions in chains, sometimes farther than the eye can reach, and the more perilous, as the navigator, though surrounded by land, can no where cast anchor. To examine these islands with greater precision, he is obliged, on account of their low situation, to come very close to them; and if a sudden gust of wind should arise, or gloomy and stormy weather set in, he is in danger of being driven against the steep coral walls, a misfortune which happened to one of Roggewein's ships. These parts have therefore been very much avoided, and, consequently, not examined by any late navigator. Commodore Byron saw, indeed, several of those low islands; and it is very probable, as I will immediately proceed to show, that some of these islands are those which were also seen by Schouten: Roggewein's discoveries, on the contrary, as they lie more south, must necessarily have escaped him. Bougainville, Wallis, and Cook, in 1773, took a much more southern course, on which a similar archipelago was discovered; the sea, however, in which the problematical discoveries of Le Maire, Schouten, and Roggewein lie, has not been explored in its whole extent from east to west. Captain Cook, in 1774, on his voyage from the Mendoza islands to Otaheite, crossed it only from N. E. to S. W. Lieutenant Kotzebue was therefore directed to look for Schouten's and Roggewein's discoveries; a commission which he executed very satisfactorily. The small size of his ship, and the very fine weather, favoured these researches; he has likewise, as his charts prove, seen more of the Coral islands, and examined them more carefully, than any of his predecessors. He has not, indeed, recognised every individual discovery made by the Dutch; but the most important doubts have been solved, as I shall now endeavour to prove.

The 26th of March, 1815, Lieutenant Kotzebue saw the island of Sales y Gomez. It has also been previously supposed, that the Gwyn rock, discovered by an American ship in the same latitude, only 5° more east, was the same with the island of Sales y Gomez; but I do not think that the situations of both the islands, according to the accounts of the Americans and Spaniards, have been examined by one and the same ship. This has been done by Lieutenant Kotzebue, who has thus placed the identity of the two islands beyond all doubt. On a very fine chart of the world, published in the year 1815, this island is still marked under the name of Gwyn, in the latitude given by the American ship, 100° 40′. According to the observations of Lieutenant Kotzebue, the island of Sales y Gomez lies in 26° 36′ 15″ S., and in 105° 34′ 28″ W.; according to the Captain of the Spanish Marine, Gomez, in 26° 28′ 47″ S., and in 105° 26′ 46″ W.

The 16th of April, Lieutenant Kotzebue discovered, in 14° 50′ S., and 138° 47′ W., a low island here and there overgrown with trees, with a lake (*lagoon*) in the middle: its length in the direction from N. W. to S. E. five miles, (sixty to a degree): they saw no inhabitants. There probably is no doubt that it is the Dog Island, discovered by Le Maire and Schouten: the best proof of its identity is its great distance from the nearest island in the west. Lieutenant Kotzebue was likewise of this opinion; but, as it did not appear improbable to him that an island resembling it might still be found there, he gave it the name of the Doubtful Island.

The 20th of April, Lieutenant Kotzebue discovered a similar, but much smaller coral island. Its length, in a direction N. N. E. and S. S. W., is three miles, and its whole circumference scarcely ten miles. It is particularly distinguished by its having no lagoon, with which almost all the Coral islands are provided. Its latitude is 14° 57′ 20″ S., and its longitude 144° 28′ 30″. Lieutenant Kotzebue, with justice, took it for a new discovery; and as it was the first, he gave it the name of the author of this princely undertaking.

Romanzoff's Island is not the island of Sondre Grund, called so by Schouten: this is inhabited, and scarcely a German mile long; it is likewise not Schouten's Waterland, for this has a lagoon; nor is it the island called by Roggewein Carlshoff; though, according to Behren's description of Roggewein's voyage, it is of the same extent; but this has a lagoon, the former not; besides this, the island of Carlshoff lies only twelve German miles east of the Shadelyk islands; Romanzoff's Island, on the contrary, above thirty miles. Schouten's course was probably too northerly, and Roggewein's course too much to the south, to see this little island, of which the Rurick lost sight at the distance of eleven miles.

The following morning an island with a lagoon was seen in 14° 41′ S., and 144° 59′ 30″ W., which was called Spiridoff; its length, in the direction of N.N.E. and S.S.W., was eleven miles. It seems to me, (though I may be mistaken,) to be the most westerly of the two islands, called by Commodore Byron King George's Islands, and whose proper name is Oura. This island, which has also a lagoon, lies, according to Cook, in 14° 37′ S., and 145° 10′ W.: he gives it an extent, in the direction N. E. and S. W., of nearly twelve miles (almost four leagues), and a breadth from three to five miles. Here every thing, therefore, coincides, latitude and longitude, extent and direction, external and internal form, by which I understand the lake in the middle of the island. But as the island of Oura lies six miles S.W. by W. from the most eastern of King George's islands, or the island of Tiookea, it might be objected, that Lieutenant Kotzebue must have seen these also; but it may easily be conceived, even without seeing his chart, that he could not see them. The island of Oura is twelve miles long;

the channel which separates it from the island of Tiookea is six miles, consequently from the S. E. point of Oura, to the S. E. point of Tiookea, is eighteen miles. Now Lieutenant Kotzebue saw only the southern end of Oura, therefore was distant at least twenty miles from Tiookea, which must have been concealed from him on account of his situation in sailing round Oura; but even were this not the case, he would not see it at a distance of eighteen or twenty miles, as these low islands, to use Lieutenant Kotzebue's own words, can be seen at the most, at a distance of fifteen miles. I should not have had the least doubt as to the identity of the islands of Spiridoff and Oura, had not Byron and Cook both found the islands inhabited, and Lieutenant Kotzebue, on the contrary, saw no inhabitants; this objection may, however, also be removed, and the more so as Kotzebue was only near the S.W. side of the island, and the residence of the islanders might be on the other side, perhaps to be nearer to the inhabitants of the neighbouring island.

The 23d of April. Land was seen on both sides; Lieutenant Kotzebue recognised that in the S.S.E. to be Cook's Palliser's islands, or, what is the same, Roggewein's Shadelyk islands: that in the S. S. W. appeared to him a new discovery. He sailed through the channel which separates the two groups, and steered to the westerly one, which forms a chain of several islands thickly overgrown with trees, and united by coral reefs. He sailed at a distance of a mile at the utmost, along the north-easterly, south-easterly, and south-west parts of this chain. From the west point, the islands took a direction to N.W. and N., but he was obliged to leave this part unexplored. The extent of the islands of Rurick's chain which he surveyed, for so Lieutenant Kotzebue called these islands to the west point, he found, exclusive of the bendings, to be forty miles: the circumference of the whole chain he estimated at least at sixty miles. The N. E. point of Rurick's chain lies in 15° 11′ S., the east point in 15° 20′, and 146° 30′ W.; the S.W. point in 15° 30′, and the west point in 15° 20′.

Rurick's chain may be looked upon as an entirely new discovery, even supposing that the eastern part of these islands is that land which Cook saw in the N. N. E., when he was at his third Palliser's Island, and called the fourth of his group. At the distance in which he was from this land, he must have considered it as an island of inconsiderable size. Fleurieu takes this fourth island to be the one called by Roggewein the Sister, and it is marked on his chart as the smallest of the Shadelyk islands.

When Lieutenant Kotzebue was sailing round the southern end of Rurick's chain, land was descried from the mast-head, but which he was obliged to leave unexplored: it lies in 15° 45′, and 146° 55′, and is, according to all probability, the island called by Roggewein Aurora, which has never been seen since its discovery. Fleurieu has calculated its situation at 15° 38′ S., and 147° 14′, which agrees very exactly with the longitude and latitude of the island seen by Kotzebue.

They had scarcely lost sight of Rurick's chain, when new land appeared in W. by S., which Lieutenant Kotzebue recognised to be the same island which is called Dean on Arrowsmith's charts. It consists, like Rurick's chain, of several islands, some of them of considerable extent, and united by coral reefs. Kotzebue sailed at a distance of a mile, at the most, along the southern side of these islands, which run in a direction from W. by N. and E. by S., to an extent of 72½ miles. Its east point lies 20′ to the west of the west point of Rurick's chain, in 15° 16′ 30″ S., and 147° 12′ W.; the west point in 15°, and 148° 22′. W. From this point the chain takes a direction to N.E.

It cannot be doubted, that this island, which Lieutenant Kotzebue called, after Arrowsmith's chart, Dean's Island*, is the same called by Byron Prince of Wales' Island, and is also marked on the charts by the name of Oanna. Byron gives its length at sixty miles in a direction from east to west, and determined its latitude at 15° S., and

* It received the name of Dean, in the year 1803, from the captain of the English ship Margaret, who would not consider it as Byron's Prince of Wales' Island, but as quite a new discovery.

the longitude of the western end 151° 53′. Deduct from this 3° 55′, which Cook found Byron's longitude of King George's Islands too westerly, and there remains 147° 58′. The longitude of this island, corrected by the astronomer Wales, is 147° 48′. The extent, direction, latitude, and longitude therefore agree exactly with those of Dean's Island seen by Lieutenant Kotzebue.

Exactly in the west, fifteen miles from the west point of this island, which Burney has made it very probable is Schouten's Vlieghen Island, Lieutenant Kotzebue, on the 24th April, again discovered land; it was of the same nature as that he had just left, that is, small islands united to each other by coral reefs. This island-chain, extending from N. N. E. to S. S. W., is fifteen miles in length, and particularly distinguished by having in the middle of the lake (with few exceptions, a property of all the coral islands,) a small island thickly overgrown with trees. He gave these islands, the middle of which lies in 15° S., and 148° 41′ W., the name of Krusenstern's islands. As Byron sailed along the north side of his Prince of Wales' Island, and then took his course to N. 82° W., it is easy to be conceived that he did not see this island, and it may therefore be looked upon as a new discovery.

Lieutenant Kotzebue, pleased to have reached the end of this labyrinth, now steered W. N. W. to look for Roggewein's Bauman's islands, in that part where Fleurieu thought that they must lie.

He, however, saw no land, and it is, perhaps, now as good as decided, that the previous opinion of the English geographers, that the Bauman islands, and Bougainville's Isles des Navigateurs, are the same, is correct; to which opinion also Captain Burney assents; however much Fleurieu tries to contradict this hypothesis. Lieutenant Kotzebue was unable to find either the islands called Roggewein by Fleurieu, which Roggewein considered to be the Cocoa and Verradër's islands, or the great islands of Tienhoven and Groningen; of the latter of which Burney is of opinion, that they are the most southern of the Solomon's islands.

The above proves that the voyage of Lieutenant Kotzebue has thrown an important light on the discoveries of Le Maire, Schouten, and Roggewein. It is only to be wished that the islands discovered by him, between the 138th and 149th degree of longitude, may soon be subjected to a second revision, and that they may furnish a harvest as satisfactory as that of Lieutenant Kotzebue. *

The 30th of April, he saw Penrhyn islands, a group of inhabited coral islands, quite similar to those he saw last. The Penrhyn islands were discovered, in the year 1788, by the English ship Lady Penrhyn, and have not, I believe, since been visited

* In my hydrographical contributions, page 173. and 189., more information may be obtained on the discoveries of Le Maire, Schouten, and Roggewein.

by any navigator. The ship Penrhyn considered the group to be only a single island, as it was only approached within eight miles, and the weather unfavourable to further examination.

Lieutenant Kotzebue found the latitude of these islands to be 9° 1′ 30″ south, and the longitude 157° 34′ 32″: 8′ more north, and 11′ more east, than they were determined by the ship Penrhyn. He counted fifteen islands, forming a circle, eleven miles and a half in diameter. The population of these islands, with respect to their size, appeared to be very considerable.

On his voyage to the north, Lieutenant Kotzebue wished to cross the chain of islands discovered by Marshall, in 1788, of which the Mulgrave islands are the southern, and Gilbert islands the northern. He could not, however, descry land between eight and ten degrees of latitude, in the longitude laid down on Arrowsmith's chart; a proof that this longitude is incorrect. The great chain of islands, which forms two large archipelagos, from 1° south to 12° north, and which bave been called, after their discoverers, Gilbert and Marshall islands, deserve a closer examination. Arrowsmith assured M. Von Kotzebue, that he had marked them on his chart according to the statements of several navigators who had seen here and there an island of this archipelago, and, consequently, he could not be answerable for their accuracy. The researches of Lieutenant Kotzebue, however, prove that the determination of the longitude of Captain Marshall is more correct than that of Captain Gilbert, to whom Arrowsmith has given the preference.

On the 21st of May, a group of low, but inhabited, islands were discovered: they were almost all united to each other by coral reefs; and reefs extended from these islands fifteen miles to the north, and twelve miles to the west. Lieutenant Kotzebue sailed all round, and found a passage of two miles broad, through which he sailed. The islands north of this channel he called Kutusoff, and the southern, Suwaroff islands. The two groups, taken together, have an almost north and south direction, and in this direction occupy an extent of twenty-five miles and a half. The channel, which separates the two groups, and is three miles and a half broad, lies in 11° 11′ 20″ north, and 190° 9′ 30″ west. Kotzebue visited these islands, for the second time, in the following year, and learnt that they were called, by the natives, Udirick, and Tagay.

As these islands are not found on any chart of the South Sea, Kotzebue justly considered them as a new discovery. An attempt has been made to dispute him the honour of this discovery, by asserting that the Kutusoffs are the same that Wallis took for the Pescadores. According to Wallis, one of the two islands seen by him lies in 11° north, and 192° 30′ west; the other in 11° 20′ north, and 192° 58′ west. The description given by

Wallis of these islands, is not in the least applicable to the Kutusoff's islands; and even supposing that this were the case, such a supposition could only be certain if a mistake of two degrees and a half, in Wallis's longitude of the Pescadores, were to be allowed. But I believe I shall be able to prove that the longitude of Captain Wallis cannot be incorrect by half a degree, or even by a quarter of a degree. East and west of the Pescadores are two points, with the situation of which we are acquainted by very exact observations, but which have likewise been determined by Wallis, by lunar observations; the true longitude of these points can, therefore, serve to prove the value of the lunar observations made on board the Dolphin. These points are the islands of Boscawen and Keppell, and the island of Tinian.

1. The 14th of August, 1767, the islands of Boscawen and Keppell were seen from the Dolphin; the 16th the ship was 1° 3′ west of them, and from the lunar distances measured that day, Mr. Wales, the astronomer, calculated the longitude of Boscawen Island 174° 7′ 39″, and that of Keppell Island 174° 10′ 24″. La Peyrouse determined the longitude of the two islands at 173° 56′ W. Taking it for granted that the longitude fixed by La Peyrouse is the true one, the error in the longitude determined by the Dolphin is less than a quarter of a degree.

2. The longitude of Tinian was determined by four series of lunar observations: two series of

lunar distances were measured in the roads of Tinian, on the 27th and 30th of September; one series on the 16th of September, when the ship was 1° 26′ to the east of Tinian, and the fourth, on leaving Tinian, 2° 15′ to the west of that island. From these four series, which, as any one may convince himself from the observations, differ very little from each other, the astronomer, Mr. Wales, calculated the longitude of the roads of Tinian 145° 55′ 25″ E. The true longitude is 145° 45′*; the error according to this was only 10′: at the islands of Boscawen and Keppell it was 13′; how then can we imagine an error of two degrees and a half, to have been committed in the Pescadores? The longitude of the Pescadores Mr. Wales has deduced from the lunar distances measured on the 27th of August, and on the 14th of September; on the first day the ship was in 5° east, and on the last day 16½° west of the Pescadores. The observations on the 27th of August, gave for the longitude of these islands 192° 27′ 30″, and on the 14th of September, 192° 51′ 52″; the mean, therefore, 192° 39′ 4″, or 167° 20′ 56″ east.† Consequently, it is quite impossible that Kutusoff's Islands and the Pescadores are the same. According to accounts which Kotzebue received on these islands,

* Espinosa Memorias sobre las Observaciones, &c. vol. ii. p. 5.

† Astronomical Observations made in the Voyages of Byron, Wallis, &c. p. 12.

there lies exactly in the west a group of islands which they call Bigini, which must indisputably be the Pescadores.

On the 1st of January, 1817, he discovered an inhabited island, low, and overgrown with wood, which, from north to south, is three miles long, and three quarters broad. Its latitude was determined at 10° 8′ 27″ N., and its longitude at 189° 4′ 46″. It received the name of New Year's Island; the natives called it Miadi. On the 4th of January, they discovered a group of islands forming an archipelago, with a basin in the middle, twenty-seven miles long, and twelve miles broad, having two passages leading to it; the one through which the Rurick sailed in was only 50 fathoms, and the one by which it sailed out, was 150 fathoms broad. Lieutenant Kotzebue remained till the 7th of February in this group of islands, which consisted of sixty-five islands, and received the name of Count Romanzoff. The latitude of the anchoring-place at the island of Otdia, the most eastern, and at the same time the largest island in this group is 9° 28′ 9″ N., the longitude, from 300 lunar distances, 189° 43′ 45″ W. This group, in a direction from west to east, occupies the space of thirty miles; its breadth is about ten miles.

A second group, likewise in a circular archipelago, was called after the late minister of the marine, Admiral Tschitschagoff; its length from N. W. to S. E. was twenty-four, its breadth only four miles.

The latitude of the middle of this group, which is called by the natives Eregup is 9° 6′; its longitude 189° 56° W.

On the 10th of February, he discovered a third group, called by the natives Kawen, which is of the same form as the two preceding. In a N. W. and S. E. direction this group is thirty-three miles long; its greatest breadth is thirteen miles. As they here found an entrance between the coral reefs, Lieutenant Kotzebue sailed into the lagoon. The largest island of this group, which received the name of Araksheef, is two miles and a half long, and three quarters of a mile broad, and lies in 8° 54′ 21″ N., and 189° 11′ W.: the south-eastern island is 8° 29′ 30″ latitude, 188° 49′ longitude.

A fourth group, consisting of thirty-two islands, and called by the natives Aur, was named after the present minister of marine, Marquis de Traversey. In a direction from N. W. to S. E. it extends thirteen miles; its greatest breadth is six miles. The latitude of the anchoring-place is 8° 18′ 42″ N., and its longitude 188° 48′ W.

A fifth group, which was discovered the 1st of March, is called by the natives Ailu; it is fifteen miles long, and five broad. Lieutenant Kotzebue found a passage through the reef which unites the islands that form the circle and incloses the basin in the middle, and sailed into it. Capenuir, the most northern island of this group, to which he gave the name of Krusenstern, lies in 10° 27′ 15″ and 190° W.

A sixth group, in 9° 51′ 30″ N. and 190° 46′ 30″, and called by the natives Ligiep, was named after Admiral Count Hayden.

These six groups, as well as the four groups of Arno, Mediuro, Mille, and Bigar, which all lie more south, but which Lieutenant Kotzebue was not able to visit, form the chain of Radack. According to the accounts of the natives, there runs to the west, and at a distance of about a degree, a second similar chain, parallel to and quite resembling the Radack chain, that is, in a direction almost from north to south. This western chain, which consists of nine large groups of islands, and three single islands, is called by the natives Ralick. There is no doubt to be entertained of its existence; for all the accounts which Lieutenant Kotzebue, while he remained near these islands, received from the islanders were found quite correct, not only in the distance, but also in the direction. The navigator, therefore, to whom the examination of this great archipelago shall be intrusted, may safely reckon on the discovery of the Ralick chain if he follows the chart which Lieutenant Kotzebue has drawn up, according to the accounts he has received, and which belong to the atlas of charts accompanying this account of his voyage. He himself missed it, as a violent current, during a stormy night, probably carried him so quickly to the west, that he passed the chain, without seeing any of the islands belonging to it. It would not have been possible for him to ply to the east against the monsoon, and against the current, without great loss of time.

I now ask, whether the groups described by Lieutenant Kotzebue are a new discovery, or whether they have been previously known? The answer is, that it is very probable that they were first seen by M. Kotzebue, which I shall now endeavour to prove. If we compare the situation of the islands of Chatham and Calvert, discovered by Capt. Marshall in 1788, on board the ship Scarborough, between the ninth and tenth degrees of latitude, and of Bass Reef-tied islands and Bishop Junction Island, so named by Bishop on board the ship Nautilus, in the year 1799, which latter are indubitably the same with Chatham and Calvert islands, it might easily be concluded that the islands discovered by Kotzebue, between the ninth and tenth degrees are identical with them. But as we know that to the west of the chain of Radack, exactly in the same latitude, there lie other groups of islands, it is, at least, just as probable, that Captain Marshall saw the western and not the eastern islands. This would, then, also be the case with the group of Ailu, which, according to Kotzebue, lies in 10° 28′, and, consequently, exactly coincides with the latitude of the islands of

Tindal and Watts, marked on Marshall's map.* That the islands of the Ralick chain, and not those of the Radack chain, were seen by the Scarborough and Nautilus seems to be probable, from the following circumstance, that Langemui, a chief of the group of Ailu told Kotzebue, that many years ago a ship with white men touched at the group of Odja, (the largest of the Ralick chain,) from which the natives procured iron in exchange; but nobody could remember a similar visit in the islands of the Radack chain. They also knew that many years ago a ship had sailed past the island of Bigini, which is the most northern group of the Ralick chain, and, as I have already said, according to all probability, the islands which Wallis took for the Pescadores. And thus it becomes probable that the discovery of the islands described by Lieutenant Kotzebue is truly owing to him, which is still more confirmed, because the island called by him New Year's Island, was seen neither by the Scarborough nor the Nautilus.

But even supposing that it could be proved that Lieutenant Kotzebue is not the first discoverer of these islands, the merit cannot be disputed him, of having first made us acquainted with their true

* In the account of the voyage of the Scarborough, no mention is made of the discovery of the islands of Tindal and Watts. They have likewise not been marked on the charts of Arrowsmith and Espinosa.

situation. This discovery was besides not accidental. What is much more meritorious than unexpectedly meeting with an island, he looked for those islands in these parts, which have been misplaced in the charts, and employed two months and a half in surveying them with the greatest accuracy. The reader has seen by his account what courage and ability he showed in minutely exploring these singularly formed islands, and what dangers he braved to penetrate into the almost hermetically closed basins. He is the first seaman who has ventured to navigate these lakes, encircled with corals; and I believe I cannot be accused of partiality when I affirm, that both in exploring these dangerous islands, and in revisiting the equally perilous islands of Schouten and Roggewein, Lieutenant Kotzebue has displayed such an undaunted courage, and such great perseverance, united with so much able seamanship, that he may, in this respect, be placed at the side of the celebrated Flinders. The discovery of the islands of Radack, so far is highly important, as it has made us acquainted with a people who are undoubtedly the mildest and most amiable in the South Sea; and I think it is not an unjust wish, that the entire investigation of this great archipelago may not be intrusted to any one but to him, who has gained in such a great degree the love and filial confidence of these good-natured people, who revere him as their benefactor, and whose return they so earnestly implored on his parting.

Lieutenant Kotzebue, on his voyage from the Radack islands to Oonalashka, touched at the islands discovered by the English frigate Cornwallis in 1807, and which are now usually marked on the charts as Cornwallis' Island. He found that they consisted of ten small islands united to each other by coral reefs. (See chart of these islands in Kotzebue's atlas.) He has determined the latitude of the middle of the group to be 14° 42′ N., and the longitude at $\frac{169° \, 3' \, 30'' \text{ E.}}{190° \, 56' \, 30'' \text{ W.}}$ I have shown in another place that these islands are probably the Gaspar Rico of the ancient Spanish navigators.

On the 20th of October, the same year, on the voyage from the Sandwich islands to Radack, Kotzebue looked about for a group of small islands, discovered on the 14th of December 1807 by the English frigate Cornwallis, and which in the charts now bear the name of Smith's Islands, after the present Captain Smith of the English navy, who has been employed for some years past in surveying the coasts of the Mediterranean, and served on board the Cornwallis in 1807, as lieutenant. Lieutenant Kotzebue has given a chart of his own of these very dangerous islands, which are surrounded for several miles into the sea with shoals, on which the Rurick was nearly wrecked. According to his observations, they lie in 169° 39′ 20″ W., and 16° 45′ 36″ N. That these islands are the same which were discovered in the year 1786 by Don Jose Camisares, pilot of the Royal Spanish Marine, on his voyage from San Blas to Manilla, I shall show in another place.

These are, probably, in a hydrographical respect, the principal results of M. Von Kotzebue's voyage. I cannot, however, pass over without noticing that his atlas contains plans of two very safe and hitherto unknown harbours in the South Sea, drawn by himself. The one is the harbour of Hana-rura, in the island of Woahoo, one of the Sandwich islands; the other is on the island of Guahon, which bears the name of La Calderona de Apura, and, according to the observations of Kotzebue, lies in 13° 26′ 41″ N., and 144° 50′ 6″ E., perfectly resembling that of Hana-rura so far, that they are both formed by reefs.

I do not think it necessary to speak of the discoveries and researches of Lieutenant Von Kotzebue in Beering's Straits; a satisfactory account of them is given in the narrative of his voyage.

KRUSENSTERN.

Asce the 30th of July, 1820.

ON THE

DISEASES OF THE CREW

DURING

THE THREE YEARS OF THE VOYAGE;

BY

FREDERICK ESCHSCHOLTZ, M.D.

PHYSICIAN TO THE SHIP.

A SHORT ACCOUNT

OF THE

PRINCIPAL SICKNESSES OF THE CREW

DURING THE THREE YEARS' VOYAGE.

It cannot, of course, be my intention to enumerate every accidental case of disease (for to do this, I need but copy my journal); but I mean to mention the effects on the health of the crew, which necessarily ensued from the external causes of disease that occurred during the course of the voyage; and here and there, when it seems not wholly unworthy of notice, add the method of cure that was adopted, and the effects of the remedies employed. My constant endeavour was, to recognise the predominant character in the several climates, which we sometimes changed very rapidly. But this was frequently rendered difficult, and sometimes impossible, either by the small number of the sick, or by the rapid change of climate, and the consequent difference of seasons; for scarcely was the knowledge of the prevailing character obtained, when it had already attained its end, and another had taken its place.

The naval physician, whose patients rapidly traverse with him extensive tracts of space, is therefore frequently deprived of the resource which is of so much use to the physician settled on shore, an acquaintance with the permanent character of the present disorders, to be able to understand the nature of new ones. Nobody will doubt, that I was not at all sorry when the cases of disease were so few, or so unimportant, that no general conclusion could be drawn from them: but I must own that, in the neighbourhood of Chili, I wished to have a patient slightly indisposed, that I might be able to investigate the cause of the changes that had taken place in one who was seriously attacked.

CHAPTER I.

VOYAGE FROM CRONSTADT TO TENERIFFE.

CATARRHS AND RHEUMATISM.

SLIGHT catarrhal fevers, diarrhœa, and rheumatic attacks, first manifested themselves in the North Sea. A sailor, who might have passed for a model of a scrophulous habit, had pimples in many parts of his body. In England his whole back was covered with small ulcers, and, on our departure from that country, they extended to all the joints of his right arm. His cure was easy and rapid: I merely applied emplast. gum. ammon. and found that it was better to let the plaster lie on two days, than to change it every day; for in this time, almost the whole rash was converted into a large blister, filled with serum, which, if it had not broken of itself, needed only a small incision, to let out the serum. The application of a salve removed the skin that formed the blister, and the red basis of the ulcers vanished.

When we were leaving Plymouth, we observed, with displeasure, that our smith was quite hoarse, without having put himself on the sick list, or given me any notice of it. During our stay at Plymouth, he had set up his forge on shore, and spent most of his time there, by which he escaped our attention. On questioning him closely, I learned that when he coughed, he felt great pain in the chest, that he expectorated very little, and that he had in particular a painful sensation in the epiglottis. The next day he told me that he had had the same complaint at a former period, (about a year before,) and had at that time spit blood. But when Captain Von Kotzebue questioned him on the subject, and reproached him for not having mentioned his illness earlier, he denied what he had said to me, and said he had spoken only of being let blood. From this, I clearly perceived that he was afraid to confess that he had come sick on board the ship, because when the sailors were selected, each was questioned respecting his health, and this man had declared himself well. Professional men will judge what hopes I had of curing a patient who was labouring under a confirmed quinsey. The pains in the chest, to remove which I was particularly anxious, in order to prevent consumption, were in fact removed in about three weeks, chiefly by the use of opium. But the pain in the epiglottis, and the hoarseness, not only remained, but continued to increase.

A great advantage, with respect to the regimen, would have been derived in this case from the voyage, which our medical books recommend, only that we cannot have the advantage of sailing always between the tropics, but must be contented with the air in the high latitudes, which is usually cold and damp. I unhappily learned the effect of the latter on our patient.

CHAP. II.

VOYAGE FROM TENERIFFE TO BRAZIL.

SPASMS OF THE LUNGS, AND BILIOUS COMPLAINTS.

In the roads of Teneriffe, a sailor was seized with such violent spasms in the lungs, as to become senseless, and gnash his teeth. On my coming on board at the third attack, I gave him a dose of *Valeriana*, with tincture of opium, after which he had no other attack.

On our entering the tropics, the effects of the heat manifested themselves; the neighbourhood of the Cape Verd islands was particularly evident. First, one of the steersmen suffered with a violent head-ache, with a bitter taste in the mouth; on which account he was continually spitting. Though, after an emetic, which operated well, some nausea still remained, he was, however, well in three days. Immediately after him, two sailors were taken at the same time, with violent cholic. Scarcely half an hour afterwards, when they had rubbed *Liniment volat.* on the most painful part, both vomited, and were at the same time taken with a diarrhœa. The following day they were well, excepting a slight degree of weakness. On this day, several sailors suffered, with pains in the stomach, and diarrhœa; to these men I gave no medicine; they were only ordered to take the soup prescribed for the sick. At the same time, others had a spasmodic affection in the larynx, with headaches, which were pretty constant; but, however, soon disappeared after the use of ether drops. Those who were not now affected with those complaints, suffered afterwards, tedious colics, and slight bilious fevers, which, however, all terminated happily.

A short time before our arrival in Brazil, a sailor was seized with very violent spasms of the lungs, which were of longer continuance with him, than with the one who had the same disorder at Teneriffe. He too was cured with *Valeriana* and opium; besides, I bathed his breast, during the fit, with ether, upon which the pain soon abated, and he fell asleep.

The smith with his quinsey was at first very well in Brazil, but the continued series of rainy days soon affected him so much, that, at the time of our departure, he could hardly swallow any thing. In the sequel, however, he got rather better.

The sailor who had had the spasms of the lungs, off Teneriffe, was once attacked by the same complaint in Brazil, after having greatly exerted himself in rowing on a hot day.

CHAP. III.

VOYAGE FROM BRAZIL TO CHILI.

RHEUMATISMS AND CATARRHS.

SOON after our departure from Brazil, rheumatisms and catarrhal and rheumatic fevers began to prevail; two of the sailors, also, had St. Anthony's fire in the legs: that of the first was soon dispersed, but that of the other, which broke out a week later, was tedious, at last suppurated, and was a long time before it would heal. A sailor who, when on shore in Brazil, had got some sore places on his feet, and did not apply to me because they were so trifling, now showed me the ulcers, covered with a blackish scab: this easily fell off, after the application of some salves; but the little ulcers would not heal. The same was the case with some cases of rash, which now began to show themselves again, turned to small ulcers, and then remained in the same stage of cure. The smallest wound, of which there were several during the storms, would not now heal, till, at length, after two days' clear weather, all became well at once.

But the smith was often so bad that he could not swallow, and grew very thin, though he had soups, chicken broth, and always drank porter. I gave him also quassia and bark.

The cook complained, one evening, of shooting pains near the liver, and had, at the same time, an incessant coughing. After taking a few drops of opium, he lost the coughing, and part of the pains; *Liniment volat.*, rubbed on the part affected, was so efficacious, that, on the following day, he had no pain remaining.

With the healing of the above-mentioned small ulcers, the rheumatic character seemed to have ceased; all our sick recovered, except the smith, who grew worse. Here it was, that I wished to have a patient slightly affected, to know to what cause I was to ascribe his getting worse. But he proceeded slowly towards his end. In Chili we endeavoured to gratify all his wishes; he had new milk to drink, and water-melons, of which he was very fond, as many as he desired. He was brought on shore, took a warm bath, of which he was very desirous, and ended, in a few days, his mortal career.

Before our arrival in Chili, a sailor had burnt his foot with boiling water.. At first it appeared to heal, but, after a continuance of damp weather, the wound suddenly became painful, and inflammation ensued. This was, indeed, soon remedied, but the ulcer that remained, healed very slowly, and caused a swelling in the knee, which soon subsided, then returned, and was not entirely removed, till we got near Kamtschatka.

In Chili, a sailor once became delirious, after

drinking some Chilian wine in the country: this soon passed over, but the following day he had severe pains in the head. The wine was probably mixed with some vegetable substance, for instance, *Lobelia Tupa*, L., to make it more intoxicating.

CHAP. IV.

VOYAGE FROM CHILI TO KAMTSCHATKA.

NO DISORDERS.

The voyage from Chili to Kamtschatka, was so fortunate for the health of the crew, that, in a period of three months and a half, none suffered from any internal complaint.

There was, in fact, no want of weather which might produce diseases; there were many days on which it rained incessantly, and the transition from a hot to a cold climate was pretty sudden and sensible. I ascribe the constant health of the crew to our great distance from high land during almost the whole of this long voyage. The low coral islands, of which we passed so many, and so near at hand, do not produce in this respect any of the effects of land: they cause no difference of temperature, either in the water or the air; they exhale no moisture, produce neither dews nor tempests, and do not impede the course of the winds.

CHAP. V.

VOYAGE FROM KAMTSCHATKA TO BEERING'S STRAITS, AND THENCE TO OONALASHKA.

CATARRHS.

After we had left Avatscha Bay, and got a little distance from land, the fog, which at this season almost always prevails near shore, did not fail to manifest its unwholesome influence. At first it produced only coughs and colds; but these were afterwards accompanied with hoarseness, and pain in the epiglottis. Slight catarrhal fevers, and St. Anthony's fire, also occurred. It was the new people whom we took with us from Kamtschatka that were chiefly affected with coughs and hoarseness. Many of them, too, complained of pains in the diaphragm, which increased on drawing breath sharply. In the course of the voyage we had frequent opportunities of observing this disorder in new people, who came on board, and who had never before served as sailors. This sensation might be considered as a symptom of *Peripneumonia catarrhalis*, if a fever took place at the same time; but I think I have observed that it proceeds from the unusual exertion of pulling the ropes, as they often hang upon them, holding by the hands, by which the muscles of the breast, and the diaphragm especially, are greatly exerted. A few days' rest easily removes this pain.

During the whole voyage in the Kamtschatka Sea and Beering's Straits, no disorders appeared except coughs, which were usually accompanied with obstinate hoarseness, and pains in the epiglottis. Hot tea made of camomile, elder flowers, and liquorice, taken in the evening, was found to be a very good remedy. Frequently half of the crew was so affected, and it did not cease till we reached Oonalashka.

CHAP. VI.

VOYAGE FROM OONALASHKA TO CALIFORNIA, AND THENCE TO THE SANDWICH ISLANDS.

RHEUMATISMS AND INFLAMMATORY COMPLAINTS.

THE very next day after we left Oonalashka, the coughs which had been suspended while we resided on shore returned; and those who had bad teeth, got the tooth-ache. All this ceased two days before the equinox; after which rheumatic pains and colics appeared, but were of no long duration.

When we entered the bay of St. Francisco, in California, we had no sick on board. Though it was now autumn in this country, and there was no reason to complain of heat, yet the more direct rays of the sun had such an effect on the two Aleutians, brought from Oonalashka, and some weak Kamtschadales, that they were ill for several days of nervous fevers; which, however, were neither violent nor of long duration.

Another Aleutian, who came on board when we were at Kamtschatka, suffered from the small itching eruption common in hot climates, with a slight fever. This eruption consists in little round blisters, (without heads,) filled with a transparent, not watery fluid, which itch only at the commencement, at which the skin of the whole part affected is reddish. At the beginning they may be removed by the use of the common salts. But if they have been neglected for some weeks, they require sudorifics.

The last day but one before we left the bay of St. Francisco, a very melancholy accident happened, which I cannot pass over in silence, though a narrative of such cases does not properly belong to the object of this report. One of the three Russians, who here entered on board the ship, a man of above fifty, a lover of the chase, set out, in the forenoon of the last day but one of our stay in the bay, with his gun, to shoot wild geese, in a pond not very far from the Presidio. In the afternoon, some natives of the country found him helpless on the ground, groaning, and weltering in his blood. They immediately gave information at the Presidio, upon which persons went out and brought him thither, and bound up with cloths the right arm, which was severely wounded, and bleeding. It was not till the evening that he got on board the ship, when we received the first news of him. About noon he had fired a shot, but at the same time a spark from the cock of the gun had fallen into a large powder-horn which he had in his hurry left open, and which hung on his right side (he was left-handed, and fired his gun with the left hand.) The powder (a whole pound) caught fire, burst the horn, and threw the man on the ground, severely wounding his right arm, and at the same time burnt his belly. The wound on the arm was not dangerous, though it was above six inches long, and four broad; all the skin, with the cellular membrane, was torn away down to the muscles, which lay quite bare, but uninjured. The patient himself said, when I first examined him, that the wound was of no consequence, I should only look at his belly, where he had great pain.

On the belly, the skin was burnt off various places in the *Regio hypogastrica*, but no deeper wound was to be seen. The patient himself was weak and pale, for he had been bleeding several hours when the wound in his arm was dressed the first time. The pain in the belly made him often uneasy. To alleviate this, I applied emollients to the part. The wound bled profusely during the night, till a tourniquet was applied towards the morning, which stopped the bleeding.

On this day we put to sea. It seemed to us cruel to leave the patient behind, as he had before been quite happy on hearing that he was to return with us to his native country; and if we left him, regret at the disappointment of his hopes might have a very injurious effect on him. Another reason for taking him with us was, that there was no physician in the country.

On this, and the three following days, the pain in the belly did not abate; the patient could not retain either meat or drink on his stomach; was weak, and began to speak confusedly: all these were symptoms of inflammation in the bowels, which was confirmed by his death, that took place suddenly, early on the morning of the 5th.

At the commencement of the monsoon, on the voyage to the Sandwich islands, the weather being rainy, and the wind variable, a sailor was seized with a shivering fit, which continued fourteen hours, after which he recovered. As the fit seized him in the night, I had given him only twice *Spiritus menth. sat.*

CHAP. VII.

VOYAGE FROM THE SANDWICH ISLANDS BY WAY OF RADACK TO OONALASHKA.

NO SICKNESSES.

Now there appears in my journal another fortunate interval of four months, in which there were no internal complaints produced by changes in the state of the air.

During our stay in the chain of coral islands, Radack, where we made frequent trips in our boats, in the lagoon, surrounded by islands and reefs, the spray of the sea-water, which was very salt, (its specific gravity was 1,0285,) and the burning sun caused the lips to swell and crack, and also produced inflammation in the eyes. Some had diarrhœa, from too frequently eating young cocoa-nuts.

I cannot here omit to remark, that on our return from one of these excursions in the boat, with a pretty fresh wind, a sailor was seized with such a chilliness, that he trembled in all his limbs, and his teeth chattered, while all the rest felt it too warm. This sailor, who was about forty years of age, had before this voyage had the bad custom, so prevalent among the common people, of getting himself blooded several times in the year. At the beginning of the voyage, on our arrival in Brazil, his usual time for being let blood was come. Both his arms, in the elbow-joints, were blue from extravasated blood, and his whole body was relaxed. Instead of being let blood, he took some cooling medicine, and the arms were rubbed a few times a day with *Spirit. Mindereri*, and in two days all the symptoms vanished. These fits returned twice, at intervals of half a year, but they were very slight. After that there were only times when he did not seem to be quite well; he was pale in the face for a few days, but did not complain of any sickness.

On the voyage to the north, from the hot climate of the coral islands, the consequent sudden change of the temperature, caused, in many, violent and continued head-aches, with pains in the stomach.

A wave, which in a storm, broke our bowsprit, broke at the same time the leg of a sailor, Peter Prishimoff, just at the calf. I mention this circumstance here, only on account of the difficulty of the dressing, and of setting the bone, which was greatly impeded by the rolling of the ship, as the storm continued twenty-four hours after the accident. It was not till the third day that the leg could be laid in splints, and then only by the considerate care of Captain Von Kotzebue, who steered another course during the dressing, that the ship might be more steady. Every violent motion of the ship gave the patient pain, especially during a second storm, which overtook us a week afterwards, as the leg always had some motion, though it was fixed and supported as much as possible. In twelve days, however, after our arrival at Oonalashka, he was able to walk about with the aid of a stick.

It is remarkable, that, notwithstanding the cold weather, in the spring, with violent winds, hail, and snow, all the crew enjoyed good health.

CHAP. VIII.

STAY AT OONALASHKA, AND VOYAGE IN THE KAMTSCHATKA SEA.

CATARRHS AND COUGHS, WITH SPITTING OF BLOOD.

During our stay at Oonalashka, at so early a season of the year as April, when we had a constant succession of rain, snow, and storms, and it was necessary to work and to live in the open part of the ship, it was no wonder that catarrhal fevers, colds, coughs, and hoarseness, should become almost general among the crew. Under such circumstances hoarseness is a complaint both difficult to cure and dangerous, because quinsey is much to be feared. A sailor who worked at the forge suffered complete hoarseness for four weeks, and even after he was cured, his voice was harsh for a whole year.

When some weeks after our arrival salmon * was caught in abundance, and the crew had it every day for dinner, frequent diarrhœas ensued, either because the cook did not always boil the fish enough, or because they ate too much of it, after having been so long deprived of fresh provisions. At first I was for treating these, like usual cases of diarrhœa, arising from catching cold, not supposing the fish to be the cause, and administered slight stimulants and stomachics, upon which, however, head-aches and sickness at the stomach followed. Upon this I gave an emetic, on which the patients were immediately better, and some stomachic medicine put an end to the whole illness in a few days. In the case of one sailor, who had a very slight attack of the disorder, I thought it unnecessary to give an emetic, and began at once with stomachics; but he was seized as the others had been, and I was obliged to have recourse to the first remedy.

Till the salmon came in, the crew received daily a sufficient quantity of fresh cod †; but nobody complained of intestinal worms, though the cod here abound in them, for which reason the

* *Salmo Lycaodon.* Pallas Zoographia Rosso-Asiatica, tom. iii. p. 370.

† *Gadus Wachna.* Pallas Zoographia Rosso-Asiatica, tom. iii. p. 182.

Aleutians, who eat these fish raw, cut them into very thin slices to avoid these animals. I mention this fact only for those who are of opinion that intestinal worms in the human species may be produced by the eating of animals, especially fish which have such worms. This is the opinion also of my estimable predecessor, Dr. Espenberg, in his instructive treatise on the diseases of the crew of the Nadeshda, during the voyage of Captain Krusenstern round the world, which, in particular, gave me so many valuable hints for the treatment of the sick on the present voyage.

It may be easily supposed that the attacks of catarrh were not confined to the sailors, but that the natives, the Aleutians, likewise suffered from hoarsenesses and coughs. The frequent rheumatisms in the joints, and particularly in the reins, and the attacks of gout, which many of the Russian fox-hunters, in the service of the Russian American Company were affected with, were generally considered as syphilitic symptoms. Among the many patients that I attended here, I could not, at this time, find any thing of a syphilitic nature. But I do not mean to say that syphilis has not yet reached this country; for how could this be possible, considering the numerous ships of various nations that have resorted hither since 1742? Nor do I know to what disorder I should ascribe a sunken nose. The patient, as he told me, had cured himself of syphilis, with fumigations of cinnabar: at present, being too much addicted to drinking rum, he was subject to gouty swellings in the arms in winter, and in the legs in summer.

Nor was any of the sailors infected with this disease: it happened only that a small blister appeared, which soon dried off on the use of *Hydrargiri præcipitati rubri.*

In June we had a great deal of fine weather, so that all the sick soon recovered, and commenced the voyage to the north in good health.

The first entrance into the foggy sea immediately occasioned coughs, as usual, but not of any consequence.

In the vicinity of St. Lawrence Island, and particularly near the ice which covered the surface of the sea on the north-east coast of the island, the cold air had suddenly so injurious an effect on the lungs of Captain Von Kotzebue, weakened by a blow on the breast-bone during the storm in April this year, that he was seized with spasms in the lungs, and fainting-fits, and when he coughed even spit blood, though in very small quantities. The distressing thought that this sickness prevented the execution of all he had proposed, and of the chief object of his voyage, had an equally injurious effect on him, for it was necessary, for the preservation of his health, to leave this cold region as speedily as possible, and to expect his recovery in repose on shore.

At the same time a sailor was seized with a violent coughing and spitting of blood, who, previous to this voyage, being on board a large ship of war, had fallen from one of the yards upon the deck, on his breast, which at the time caused him to spit blood. During this voyage, too, he often suffered with pains in the breast, which it was necessary to relieve by weakening remedies. From this time he was excused from all hard labour.

During the voyage from St. Lawrence Island to Oonalashka rheumatisms were very frequent.

CHAP. IX.

VOYAGE FROM OONALASHKA TO THE SANDWICH ISLANDS, AND THENCE BY WAY OF RADACK TO THE MARIAN ISLANDS.

SCIATICA AND INFLAMMATORY COMPLAINTS.

On our arrival at Oonalashka, three sailors were attacked by sciatica, which at first consisted in pains along the ischiadick nerve; after a time the pain left this place, and settled in the calf, which grew hard and drew the leg crooked, the common sinew of the *M. Psoas magni*, and of the *Iliaci interni*, contracting and becoming hard. The use of internal remedies, and the external application of stimulant salves, terminated the complaint with a serous eruption, which covered the whole calf and the knee: this eruption soon healed.

On the voyage to the Sandwich islands two Aleutians and some sailors suffered from the heat. But the whole sickness consisted only of head-ache and a general lassitude, and was soon removed by *Cremor tartari.*

From the Sandwich islands to Radack we had some very hot days, from which many of our people, by frequent perspiration and bathing in the very salt sea-water, were affected with a small, red, and very itching eruption in different parts of the body.

Near the coral islands we had frequent rains and gusts of wind, on which a sailor was attacked with rheumatism in both legs, and another with sciatica. This sciatica was at first attended only with pains in the knee-joint, but it afterwards changed into a general rheumatism. The sinew of the *M. Psoas magni,* and of the *Iliaci interni,* which was at first contracted and hard, gradually yielded to the application of salves: but small hard swellings appeared on the sole of the foot, and the lower side of the fore-arm, which were only painful at their origin. The patient had got this rheumatism by bathing on the bowsprit, where he let the water on his body be dried off by the wind.

Soon after, two other sailors were seized with rheumatism in the reins, which was very painful, because every bend of the body caused pain. They had contracted this disorder in the same manner as the preceding.

CHAP. X.

VOYAGE FROM THE MARIANA ISLANDS TO THE PHILIPPINES, AND STAY AT LUÇON.

BILIOUS DISORDERS AND VACCINATION.

A SHORT time after we left Guahon, colics began to prevail, as they had done in the neighbourhood of the Cape Verd islands. In some the colic changed to diarrhœa, and was then soon over: others were attacked with head-aches, sickness at the stomach, and gripes, and frequent retchings. In this case emetics are of no avail; but the common salts are the best remedy.

Near Luçon, in the Chinese Sea, catarrhal fevers and rheumatism manifested themselves on account of the great dews which often fell even before sunset.

On arriving in the bay of Manilla, our first business was to enquire respecting the small-pox in this island, for the five Aleutians on board our ship had not had that disease, as indeed it has not yet penetrated to the Russian American possessions. One of the Kamtschadales likewise begged to be inoculated, as he had never had the small-pox, though he had twice undergone the operation. A physician of Cavite, who inoculated with the cow-pox once a week in the neighbourhood, sent us two Tagalese children which had the vaccine. As the progress of this disorder was different from that of vaccination in Europe, I think it not superfluous to relate it. According to the course it took here, we should take it for spurious; but in its nature it was very near the genuine. The Spanish physician assured me, that the vaccine with which he inoculated the Tagalese had the same stages as in Europe.

In four of the Aleutians and the Kamtschadale I made only one puncture, but in the fifth Aleutian two, because I could not depend on the first. On the second day, the places inoculated were merely a little red: on the third day they were rather redder. In three of the Aleutians these places had already risen a little, and assumed a yellowish colour, on the edge of which there was a little raised pimple, which seemed to be filled with matter. These three patients were always a day in advance of the others, and one of them felt already on this day a prickling pain through the arm up to the shoulder. On the fourth day, the inoculated places on the three Aleutians formed elevated, longish, and cornered figures, which contained matter. In the other two Aleutians, the places were only a little elevated and yellow; in the Kamtschadale only red. On the fifth day, the elevated places in all the Aleutians were already suppurated; the middle sunk in, and covered with a little reddish crust in the form and size of the incision. The three first Aleu tians had pains in the axillary glands, and one of them also head-ache. In the Kamtschadale the redness round the puncture had also vanished.

On the sixth day, there was on the Aleutians, round the pox, a raised red rim two lines broad, the axillary glands were swelled in all of them. On this day I again inoculated the Kamtschadale in two places, the first being entirely healed up. In one only of the Aleutians, the youngest, the lymph was still quite fit for inoculation. But this time, also, I foresaw that it was in vain, because the lancet would not pass easily under the skin. The event was, in fact, the same as the first time, so that I dismissed him in three days.

At the same time, I inoculated another Kamtschadale with the vaccine who had likewise been twice inoculated in Kamtschatka without effect. Three days after, the places vaccinated had risen considerably, were red, and caused an itching; but this was the highest point to which they advanced, for, on the following day, all the symptoms had disappeared, and did not return. He made me easy by telling me that the two preceding attempts to inoculate him in Kamtschatka had terminated in the same manner.

On the seventh day, the red circle round the places on the Aleutians had spread farther, and the part all round was hard. On the eighth day, the circle was diminished both in size and redness; on the ninth, the circle and the swelling of the axillary glands had considerably abated, a copper-coloured scab had formed over the pox in all of them. From that day the scab daily increased, and the red circle diminished. On the thirteenth day, the skin of the red circle in all of them came off, and the edges of the scabs separated. On the fifteenth and sixteenth days all the scabs fell off. On the nineteenth day all were quite well.

Many of the sailors suffered so much by the heat in the hold while unloading the ship, that they had very severe colics.

Ulcus syphiliticum in glande duobus præteritis diebus comitatum ulceribus frontalibus angustis semicircularibus lichenoideis ; omnia hydrargyro curata expulsataque, etsi per longum usitato.

CHAP. XI.

VOYAGE FROM LUÇON TO THE CAPE OF GOOD HOPE.

NERVOUS FEVERS, MEASLES, FLUXUS PANCREATICUS.

IMMEDIATELY after we left Manilla Bay, the heavy night-dews in the neighbourhood of the high land occasioned catarrhal and rheumatic fevers.

The crossing of the line near or between the East Indian islands had this time no bad effect on the crew; they did not even suffer from the heat, which is probably to be ascribed chiefly to the fresh north wind, and the absence of calms.

After the departure from the tropic of Capricorn, and the loss of the south monsoon, the first cool south wind, with a change of the temperature from 23° to 18° Reaumur, caused various sicknesses; every day some sailors were attacked with fevers, accompanied with severe head-aches and great debility. The debility was frequently much increased by diarrhœas, which besides sometimes occurred alone.

At the same time the two youngest Aleutians got the measles, the one two days earlier than the other. First they had much pain in the head and neck, and then in the gums and palate; these symptoms abated on the eruption of blackish violet-coloured spots over the whole body. The spots, which in the Europeans are red, were of a dark violet in the Aleutians, who have a blackish skin. The thin crust separated on the second and third day. The first patient recovered on the third, the second on the fourth day of the disorder.

A sailor, (a Tartar,) who first suffered from a head-ache, was seized in two days with diarrhœa, with discharges of blood, which continued a day and a half; after which a slight diarrhœa remained. The pain in the head abated at the same time; but two days after, he was seized with a strong fever, gripes, with increased diarrhœa, and pains in the stomach. The next day he was very weak, had tenesmus, and discharges of blood mixed with slime. In the afternoon, the fever returned with increased gripes, and continued the whole night through, with frequent and violent tenesmus. On the third day the same symptoms returned, but in rather a less degree. The food he took caused sickness at the stomach, and vomiting. I took this disease for dysentery, (*Fluxus cœliacus,*) and gave stimulants and analeptics, which did not help the disorder, but made it worse. The *Linimentum volatile,* which was at the same time rubbed on the belly, had a better effect, because it acted as a purgative. I resolved, therefore, to give the patient tamarind-fruit, as a gentle purgative, upon which he grew visibly better, only he was very weak; but as soon as I attempted to give him any thing strengthening, he was again as ill after the first small dose as he had been a week before. I, therefore, continued the tamarinds, only giving them less and less frequently. On the 21st of March, at the hour of full moon, he felt a stitch in the side, and from that day grew better. He was still weak, and his appetite bad, when we arrived in Table Bay.

CHAP. XII.

VOYAGE FROM THE CAPE OF GOOD HOPE TO CRONSTADT.

FLUXUS PANCREALIS, BILIOUS FEVER, CATARRH.

In Table Bay, during a high wind, with very dry air, the Tartar was again seized with pain and hardness in the stomach, with sickness and vomiting, and tenesmus, accompanied with strong discharges, mixed with blood. At the same time, another sailor, (the same who had had the cough with spitting of blood in the Kamtschatka Sea,) was attacked with the same disease, which I cannot but consider as *Flux pancreatis.* Bitters and stomachics only increased the evil; I, therefore, had recourse to the tamarind-fruit, upon which the hardness of the stomach daily lessened, as well as the sickness, and the tenesmus soon ceased. The last sailor was so well at the commencement of the S.E. monsoon, that he took stomachic medicines without injury, and was soon entirely recovered.

The Tartar, however, could neither get rid of his sickness at the stomach, nor bear strengthening remedies; and, therefore, as he was still weak when we approached St. Helena, the neighbourhood of land had a very bad effect on him. The island of Ascension had a similar influence on him; on both occasions he experienced for one day before, and one day after, increased sickness, and pains in the stomach. He then improved gradually till the end of the S. E. monsoon, which brought us only to the equator. From this time he grew worse: he threw up a maw-worm. I gave him a purge to relieve him from the slime which had accumulated in the lower belly. As he grew rather better after this, I continued to give him tamarinds till he reached the latitude of the Cape Verd islands, where the air grew better, as the N.E. monsoon no longer blew so near from shore, and the patient visibly improved: the full moon, which happened at this time, perhaps also contributed to his rapid improvement. Some days later, when he had no more sickness at the stomach, and had a twitching all over the stomach, (which I had formerly observed in him to be a symptom of getting better,) I gave him *Tinctura valerianæ ammon.* with a drop of *ol. menthæ pip.* which agreed very well with him. As he had previously been used to sickness at the stomach, when he had appetite, he now received a glass of wine daily an hour before dinner. Bark and bitters at length put an end to his three months' illness *; but, for some weeks, he could not bear to tread hard on the

* During these three months we were constantly in the neighbourhood of Africa, from Madagascar to the Cape Verd islands.

ground, because every violent motion of the body gave him a pain in the left side.

During the N.E. monsoon, to the south of the Cape Verd islands, some sailors were affected with constant sickness at the stomach, and retchings. Emetics were of no avail; but purgatives put an end to the sickness. One of these patients had vomitings for seven days, so that I was often obliged to allay the vomiting by medicines.

In the English channel, many were affected with hoarseness, without coughs, others with diarrhœa;

In conclusion, I will only add, that during the three years of the voyage,

1. No contagious disorder prevailed;

2. That we never needed fumigations with nitric acid, because the crew were always in the upper airy hold; and,

3. That we never had any sign of scurvy; the honour of which is due to the planners and to the conductor of this expedition.

REMARKS AND OPINIONS

OF THE

NATURALIST OF THE EXPEDITION,

ADELBERT VON CHAMISSO,

DOCTOR OF PHILOSOPHY, AND MEMBER OF THE IMPERIAL LEOPOLDINE ACADEMY OF SCIENCES, AND OF THE SOCIETIES OF NATURALISTS, AT BERLIN, MOSCOW, LEIPSIG, &c.

Τὸ τοῦ πόλου αστρον.

The Naturalist of the expedition has been expressly commissioned to draw up the following papers, which, from the nature of the case, are to contain researches, observations, corrections, and discoveries, in which each of the members of the expedition had a share, and which are to be considered as the fruit of their joint labours. The author desires, expressly, to guard against the suspicion of desiring to attribute to himself the merit of others.

On the other hand, he will be alone answerable for the composition, and for the opinions which he expresses, which every one, perhaps, would not participate with him.

He farther recognizes only the German text: for the various foreign subjects of which he had to treat have made him too sensible how difficult it is, when aiming at brevity to avoid obscurity, for him to answer for translations of which he cannot judge.

ADELBERT V. CHAMISSO.

Berlin, Dec. 1819.

A VIEW OF THE GREAT OCEAN, OF ITS ISLANDS, AND ITS COASTS.

On the western side of the Great Ocean, a series of islands and peninsulas form a bulwark before the much indented coasts of the continent. Behind this bulwark lies New Holland, as the south-eastern point of the mass of land of the Old World. The connection of the countries is interrupted, between New Holland and Asia, by various passages, but may be easily restored in the imagination; and thus the island of Borneo, which must otherwise have been considered as a continent by itself, appears in a natural connection.

The Indian Ocean penetrates from the open South Sea, between the two promontories of our world, Africa, and New Holland, forming a spacious basin.

We return to the Great Ocean, which has been called the Pacific Ocean and the South Sea, two names which are both equally inappropriate.

The Philippines form its coast to the extreme west, between the equator and the tropic of Cancer; they lie before the masses of land which we consider as continuations of the continent, and are connected with them, especially with Borneo, by intermediate islands, and groups of islands.

From Magindanao, the most southern of the Philippines, the chain of forelands stretches to the south-east over the Moluccas and Gilolo, New Guinea, and various adjacent archipelagos, as far as to New Caledonia, and the New Hebrides lying before it, under the southern tropic. The separate mass of New Zealand may be considered as the southern end of this bulwark, and Norfolk Island indicates its connection. Beginning at Luçon, the most northerly of the Philippines, the chain of forelands extends towards the N.E., over Formosa, smaller groups of islands, Japan, the Kuriles, the Kamtschatkan peninsula, the Aleutian Islands, the American peninsula of Alashka, and connects itself with the continent of the New World, in the sixtieth degree of north latitude.

Burning volcanoes rise every where along this coast, at least from the New Hebrides to the continent of America. In New Zealand, also, volcanic productions have been discovered, and shocks of earthquakes perceived. Further inland, along the coast here described, volcanoes occur only in isolated places. It is to be observed, that the earthquakes, which shake the Philippine islands, are not felt in the island of Paragua (the Patavan of the English charts), which lies S. W. of Luçon, between Mindoro and the north point of Borneo.

The western coast of the two Americas forms the eastern coast of the Great Ocean. It runs clear and uninterrupted, broken only in the extreme north and south, by a few islands, and has only one inlet, the gulph of California, towards the northern tropic.

A burning volcano rises on the sea-shore in New California, and the peninsula shows a volcanic nature. The lofty back of the New World, which is turned to the Great Ocean, presents a series of burning volcanoes from New Spain to the southern point of America.

The islands of this watery basin, thus bounded, are divided into two chief provinces, and a detached group.

To the first province belong the islands which lie to the east of the Philippines, between the equator and the northern tropic, as far as to the meridian of Greenwich. The second province lies southwards of the line, towards the tropic, which it passes over in some places, and stretches from west to east, from the Forelands to Easter Island, and the Rock of Sales y Gomez to the extent of above 100° of longitude. The group of the Sandwich islands lies detached towards the northern tropic. The islands of the second province, the Sandwich islands, and New Zealand, are to be united with respect to the people who inhabit them.

In a geognostic view, these islands belong to two different formations. The high islands, which are the smaller number in the Great Ocean, though they form the principal groups, are universally, as in other seas, and especially in the Atlantic Ocean, of a volcanic nature. The Mariana islands form, in the first province, a mountain chain, running parallel with the Philippines, and which might be compared with the Forelands that bound the ocean basin; it contains, like them, especially towards the north, volcanoes that are still burning, whereas the islands that lie separate, in the middle of the basin, seem to be for the most part extinguished. A volcano burns at Tofua, in the west of the second province; and Mouna Wororai, in Owhyee of the Sandwich islands, discharged a stream of lava by a lateral eruption, so late as the year 1801. In the Friendly and Marquesas islands primitive kinds of rocks occur; in Woahoo we found only porphyry and almond-stone.

The low islands, that is, the coral islands, and reefs as they are called, present us with a quite peculiar formation, which we had no want of favourable opportunity to examine, and which we more particularly describe in our paper upon Radack, from which our knowledge and observations on this subject were chiefly derived.

These islands, and circular island-groups, are table mountains, which rise perpendicularly from the depths of the ocean, and near which the lead finds no bottom. The surface of the table is below water; only a broad dam round the circumference of it (the reef), reaches the surface at low water, and bears on its ridge or back, the sand banks, (the islands,) which the sea throws up, especially on the windward side, and at the projecting angles of the circumference. The reefs and islands, therefore, inclose an internal basin, or lagoon. It is only when the extent of the *plateau* is very small, that the lagoon is filled, in which case a single island is formed instead of a group of islands. As far as the dam can be examined, it consists of horizontal layers of a limestone (congeries) formed of coral sand, or fragments of madrepores. Masses of rock, often measuring a fathom, thrown upon the dam, are of the same stone, which often contains only larger fragments of madrepores, than the layers exposed to view above; and we are of opinion, that the whole construction, the table or *plateau* which forms the basis of the group of islands, consists of this same kind of stone. It is a species of rock of new formation, and which still continue to be produced. This same kind of stone, or rock, lies under the same climate, at the foot of all high islands, at least in places, and forms the coral reefs, by which many of them are entirely surrounded.

The plains at the foot of such mountains, which form the edge or border of the islands, appear to be similar reefs, which the sea, that has formed them, formerly covered when it was higher than it now is. These coral reefs, lying close to high land, sink sloping into the sea, so that the waves, rolling over an inclined plane, foam in surf, and do not dash and break, as against the upper ledge of a wall of rock. * It is the same stone, in which, on the coast of Guadaloupe, petrefied human skeletons are found inclosed. We have seen the celebrated specimen in the British Museum, and had an opportunity of accurately comparing the stone in the Berlin mineralogical museum. †

These coral reefs, low island-groups, and islands, are extremely frequent in the Great Ocean, between the tropics, and especially between the limits assigned above to the two island provinces. Sometimes we found them singly; sometimes in rows, which seem to indicate a ridge in the bottom of the sea; sometimes in the vicinity of the high

* We observed this with particular accuracy in Woahoo, between Hana-rura and Pearl River, where we passed in one of the boats of the natives, along the reefs, and several times backwards and forwards through the surf. On the outside of it, some boats were engaged in fishing in three or four fathoms' water.

† We met with this reef-stone, in the year 1817, at Owhyee, at the foot of the lava, which flowed in 1801 from the mountain of Wororai, and where there is no proper reef. Here it contains fragments of lava; otherwise it is identical with that collected on the low islands. The stone of Guadaloupe is precisely one and the same with the fine-grained varieties of it. We found this reef-stone, and in some places reefs, at Manilla and Guahon. With respect to the varieties, composed of larger fragments, some local differences may arise, from the difference of the species of madrepores, of which they are chiefly composed. We are of opinion that the species which live on the spot, furnish the elements for the stone which is formed.

islands, and, as it were, annexed to the groups which they form. Their formation, however, is not exclusively peculiar to this basin. The ill-famed sea between the coast of New Holland, and the series of Forelands, beginning at New Caledonia, and extending beyond Torres' Straits, (the sea where La Peyrouse perished, and Flinders narrowly escaped the same fate), is filled with reefs and banks of this kind. In the Indian Ocean lie many (for the most part uninhabited) islands and reefs, which belong to the same formation. Such are the Chagos, Juan de Nova, Cosmoledo, Assumption, the Amirante, the Cocos islands, &c. The Laccadive and Maldive islands, as far as we may venture to infer, from the imperfect and unsatisfactory accounts that we have of them, may probably be included; and, lastly, the stone of Guadaloupe, shows us the elements of this formation in the Atlantic Ocean, though in that narrow strait it has not become an independent production of new land.

In the Great Ocean, and in the Indian Sea, the high and low islands lie towards the west, leaning, as it were, against the adjacent continents, which are all variously indented, from the east towards the west, and we may repeat the same observation, only on a more limited scale, in the Atlantic Ocean. The Gulf of Mexico represents, in the most striking manner, the Chinese Sea, and the archipelageos that bound it. The island of Borneo may be compared with Yucatan; only between Timor and Cape Van Diemen in New Holland, the isthmus is broken through, which corresponds with the isthmus of Darien in America.

On the western coast of the Old World, Europe, with the Baltic, the Mediterranean, and the adjacent islands and peninsulas, forms the only exception to the law which results from the contemplation of the terrestrial globe.

Though on the coral reefs, and the kind of rock of which they consist, we no where recognize the skeletons of the Lithophytes in their original places, and the spot where they lived and grew, and in this differ from Flinders, for whose observations we have otherwise the highest respect*, we cannot but believe, that in those parts of the sea where the enormous masses of this formation rise, even in the cold and unillumined bottom of the ocean, animals are continually employed, in producing, by the process of their life, the materials

* In his voyage, *passim*, he assumes, that the skeletons of the madrepores are converted into reef-stone on the spot where they grew, by the filling up of their interstices with coral sand, or the fragments of other madrepores, while their upper branches continue to grow, and others build on the ground thus raised. Forster touches this subject but slightly, and what he says of it is not worth notice. The supposition that *polypi* producing limestone, live only in the walls of the already-existing reefs, and their internal lagoons, would not explain the first origin of these reefs, the perpendicular height of which cannot well be taken at less than a hundred fathoms.

for its indisputably continued growth and increase*; and the ocean between the tropics seems to us to be a great chemical laboratory of nature, where she confides an important office in the system of her economy to these imperfectly-organized animals that produce lime-stone. Objects, it is true, appear magnified in proportion as the eye is near to them, and he who, in the midst of these islands, contemplates their formation, may be disposed to assign to it greater importance in the history of the earth, than the reality justifies. An accurate description of the state of these reefs at different periods, for instance, at an interval of half a century, if it were possible and really undertaken, must contribute to throw light upon many points of natural history.

It is to be observed, that the low and small points of land which belong to this formation, have no influence on the atmosphere. The constant wind sweeps over them unchanged, as over the unbroken surface of the waters. They cause no precipitation of water or dew, and though we paid the greatest attention, we never observed in them the phenomenon of the mirage, which their flat profile is peculiarly well adapted to render evident to the eye. As an exception from this

* Captain Ross found in Possession Bay, in 73° 39′ north latitude, living worms in the mud, which he drew up from a depth of one thousand fathoms, and the temperature of which was below the freezing point.

rule, we must mention the thunder-storm which hung over the large and palm-covered Penrhyn's islands at the time we saw them.

In the Sunda islands, organized nature, by the abundance and luxuriance, the grandeur and variety of its productions, fully answers the expectations which we form of a continent lying under the equator. Unfortunately, it is hitherto but little known. Since the time of Rumpf and Bontires, only hasty travellers have contemplated it with a scientific eye, and it is but now that men of learning and collectors are repairing from different quarters to reap the ample harvest. It is assimilated to the nature of the south of Asia, from which, however, it is distinguished by many peculiarities. New Holland seems to us to present a creation of its own, which declines to receive accessions of wealth from the adjacent countries. Organized nature has evidently spread from the Terra Firma to the Forelands and islands, that is, contrary to the course of the winds, from west to east, over the points of land that rise from the Great Ocean.

The appearance of nature in the eastern islands of the South Sea, reminds us at once of Southern Asia and of New Holland, and is wholly dissimilar to America. Many genera of plants are diffused over the Indian and the Great Ocean from the African coast to these islands, which are sought in vain on the opposite shores of America.

On the high island of Pascha (Easter Island), which lies next to this coast, and apart from the others, Forster (besides the cultivated useful plants which have followed the progress of population from the west hither) has enumerated only nine species growing wild.

Forster found in New Caledonia three American plants, viz. *Murucuia aurantia*, *Ximenesia encelioides*, and *Waltheria Americana*. To these we may add some widely-diffused species, chiefly strand-plants. *Ipomæa maritima*, *Dodonæa viscosa*, *Suriana maritima*, *Guilandina Bunduc*, all of which we found, with others, at Radack, *Portulaca oleracea*, which we found on Romanzoff, &c.: but what do these prove against the testimony of the whole vegetable kingdom? We select, by way of example, some distinguished characteristic species.

The fifteen kinds of *Dracæna* with which we are acquainted (*Dracæna borealis* is *Convalaria*, Pursch.) are scattered from the eastern coast and south point of Africa, over India and the islands of the Indian and the Great Ocean. None of them occur in New Holland; two of them are found in New Zealand; and *D. Terminalis* is universally spread from India to the eastern islands of the Great Ocean. Twelve kinds of *Amomum*, (there is besides one in Jamaica peculiar to that island,) and both the *Curcuma* are spread over the same tract; and the species which grow on the mountains of the Sandwich islands are likewise natives of India. These species do not occur in New Holland.

Of the genus pandanus, one kind is found in Africa, one in Arabia, and one in the Mauritius. Brown has reckoned two kinds in New Holland, we four or five in Luçon, two or three in Guahon, and one kind is spread over all the islands of the Great Ocean. A plant related to this species is found in Norfolk Island, (F. *Bauer* in Brown's Prodromus, 341.) and in Woahoo.

One sago-palm grows in Madagascar, the other kind in the islands of the Malayan archipelago, and in the Philippines. The cocoa-palm does not pass Torres' Straits, and is not found in New Holland. The *Tacca pinnatifida* is a native of Southern Asia, New Holland, and the islands of the Great Ocean. The *Phormium tenax* is found only in New Zealand and Norfolk islands. The *Baringtonia speciosa* belongs to the coast of Asia and the islands of the Great Ocean. Two kinds of *Aleurites* occur in the Molucca islands; a third kind constitutes a chief part of the vegetation of the South Sea islands. One kind of *Casuarina* is found in Africa, another in India, and the islands of the Great Ocean; the others are natives of New Holland exclusively.

Of the New Holland numerous species of *Metrofideros*, *Melaleuca*, and *Leptospermum*, only one is found in India, several in New Zealand, New Caledonia, Otaheite, and the Sandwich islands;

the species *Eucalyptus* seems to be confined to New Holland. Of the New Holland form of leafless acacias, one kind is found in the Mauritius, and one in Cochin-China. In the Sandwich islands such a one is the pride of the forests, and the principal tree. The *Santalum* (Sanders-tree) an Indian genus, to which Brown has discovered five new species in New Holland, is found in the Fidje and Sandwich islands.

We confine ourselves here to these few features. The predominant families of plants in Luçon are the *Urticeæ*, the *Leguminosæ* of various forms, the *Contortæ* and *Rubiaceæ*. We counted twelve species of palms, and there may be some others, but they are only subordinate. *Nipa* remains hid in the morasses; other dwarf species in the obscurity of the woods of fig-trees; and only the cocoa-tree, which, where it is planted, forms fine woods, answers the expectation which this kind of plant excites.* The most beautiful of grasses, the bamboo-cane, of which there are several species already characterized by Loureiro (*Flora Cochin-Chinensis*), gives the landscape a peculiar and pleasing character.

This rich Flora seems to become more scanty in the islands of the Great Ocean, from the west towards the east. The palms vanish first, except the cocoa, which seems to belong to the lower islands, and overshades the Penrhyns, for instance,

* On the beautifully verdant shores of Gaspar and Sunda Straits, also, we no where found the palm species predominant.

with an airy canopy, below which the eye sees the sky shining between the slender stems: the bamboo retires; the other elements of the Flora are otherwise combined. Otaheite has many plants which seem to be wanting in the Sandwich islands, and these again, others that are not found in Otaheite.*

The heights of Owhyee, adjacent to the regions of eternal snow, remain the most mysterious, and the most tempting task for the botanist, so long as the harvest which Menzies reaped there shall be withheld from the learned world.

Easter Island, the nearest to the American coast, is the most scantily apportioned, but indeed it is situated beyond the tropic.

Assumption (an inhospitable volcano to the north of the Ladrones, in about 20° north latitude) afforded a richer harvest to the scientific gentlemen of La Peyrouse's expedition.

Vegetation seems to settle late and hesitatingly in the lower islands. Numerous sand-banks of considerable extent, shine, white and bare, above the surface of the waves. When once commenced, it perhaps makes a rapid progress, yet it appears in a very unequal state of forwardness in the several islands and island-groups.

Where the cocoa-tree has penetrated, the earth is ready for the reception of man, and man is

* In Otaheite the *Baringtonia speciosa* and *Casuarina equisetifolia*: in the Sandwich islands the *Santalum*.

seldom wanting in the South Sea islands, where he can find subsistence.

The Fauna of the Sunda islands presents us, for the most part, with the same families and genera that are natives of Southern Asia, but many of the species are peculiar to them.

Among a numerous variety of apes, we distinguish the ouran-outang, the species most similar to man, and the nearest relation to which is found in Africa. We found also the Asiatic elephant, a peculiar species of the rhinoceros, several stags, swine, &c.

The *Mammalia*, found in New Holland, have almost throughout shown new genera and species, and new and singular forms. The kangaroo, the largest of the species yet examined, is but small, compared with the animals of the other continents, but the existence of larger hitherto unknown species, is certified by the testimony of several travellers. The difference between the birds of the two countries is less striking. Of two kinds of cassiowary, one is found on the Sunda islands, the other in New Holland.

The islands are the richest in birds: parrots, fowls, and pigeons, and the genus *buceros*, are the most remarkable.

The *Psittacus formosus* and the *Menura* are two kinds peculiar to New Holland. The birds of paradise seem to belong exclusively to New Guinea, a country with which we are at present so imperfectly acquainted.

The islands and the continent, according to the climate under which they lie, are equally rich in the larger *amphibia*, and the crocodile in particular is found in both.

Several kinds of animals have spread from the northern point of Borneo to the adjacent islands. In Jolo (Sooloo in the English map) we find the elephant, and in Magindanao several of the larger kinds of monkies. Fewer *Mammalia* have passed from the northern point of the same island to Paragua, and in Luçon, the most northern of the Philippines, the number of kinds is very limited.

The great bat (*Vespertilio Vampyrus*) has spread to the most westerly islands, in the northern province to the Mariana islands, and in the southern to the Friendly isles. A small kind occurs in the Sandwich islands. The animal of the class *Mammalia*, which is the most widely spread, is a rat, which is met with every where, even on Easter Island.

The land-birds are pretty numerous and various in the high islands, and many kinds of them seem to have no other country.

A kind of crocodile has spread as far as the Pelew islands. At Eap such an animal has been but once seen, and in the southern province in the Fidje islands. (Mariner's Tonga, vol. i. p. 327.) An ignon is found further on, even to the Ladrones and Eap.

All the islands are extremely poor in insects. It is remarkable that the flea had not followed the dog and man to the islands of the Great Ocean, and that it was first introduced there by the Europeans. According to our experience, this remark holds good of the islands of the first province, as well as of New Zealand and the Sandwich Islands.

The common worm seems to be an universally diffused animal; we found it in the low islands, wherever mould had been formed.

We now proceed from the view of inferior nature to the contemplation of man.

The first race of men that attract our attention are the Papuas, or Austral negroes, with woolly hair, projecting jaw-bones, thick lips, and black skin. These negroes appear to us as the aborigines of the East Indian islands, and of part of the nearest continents and forelands before the immigration of other nations, and the commencement of history. They have been expelled from most points by foreign nations who have immigrated, and have fled before them into the mountains of the interior, which they inhabit as separate savage tribes.

We meet first in the west, in the island of Madagascar, as in the East Indian islands, with two decidedly different races of men. The Madecasses, who are the most known to us, and who, divided into separate hostile kingdoms, possess all the coasts, are one people, and speak one language. Drury calls them also negroes. Their hair is long and smooth; some families of princes are distinguished by a fairer complexion. Their resemblance with the Malay race, and, in their language, the identity of many roots with the other dialects, are remarkable. The influence of Mahometanism in their manners is evident. They have always had a commercial intercourse with the Arabs. The Vinzimbers, with almost woolly hair, and skulls artificially moulded, with peculiar manners and languages, seem, though now dispersed and unsettled, to have been the aboriginal inhabitants of the island.

Shall we derive the Madecasses from India, and the Vinzimbers from Africa? or shall we unite them with the Papuas, to whom they may be compared? *

The smaller islands of the Indian Sea were uninhabited before the Europeans came.

We recognise the Austral negroes in the aborigines of Cochin China, the Moys, or Moyes, who, about the beginning of the fifteenth century, were driven by emigrants from Tonquin, of a Tartar race, into the mountains between Cochin China and Cambogia, their present abode†, and in the mountaineers of the Malayan peninsula, who are called Samang Bila, and in the southern part

* We have made use principally of "Madagascar, or Robert Drury's Journal," London, 1729: his Vocabulary, and that of Hieronimus Megisserus, Leipsig, 1723.

† Chapman, in the Asiatic Journal.

Dayack. The people of the Andaman islands are evidently of the same race. The Papuas still exist under different names, in the interior of several of the Malayan islands, and it appears that they formerly lived on all. They are frequently mentioned in the early Arabian Voyages.*

The Aetas, or Negritos del Monte, the Papuas of the interior of the Philippine islands, are likewise aboriginal inhabitants of this archipelago, Los Indios of the Spaniards. Those who are whiter, are foreign conquerors; and the names of places which still exist along the coast in the languages of the Papuas, are monuments which they have left behind them of their right of possession. We find the same race of men again under similar circumstances at Formosa; and the history of Japan mentions black inhabitants, who have been met with on the islands of the southern coast of Niphon.†

We find the Austral negroes in almost undisturbed and undivided possession of New Guinea, or the land of the Papuas, and the islands lying more to the east, which, with the New Hebrides, and New Caledonia, form the chain of forelands; and we recognise them in the tribes which Forster enumerates in his second chief race of the South Sea islanders.‡

* J. Leyden, Asiatic Researches, vol. x. p. 218.

† Mithridates, First Part, p. 569.

‡ J. R. Forster, Observations, p. 238.

They exist on several of the most eastern of these islands simultaneously with the other chief race, and appear in many places very much changed by mixing with it.

Crozet in "Nouveau Voyage à la Mer du Sud," found these negroes among the inhabitants of the north point of New Zealand, where later travellers have not seen them.

The west coast of New Holland and Van Diemen's Land are inhabited by real Papuas, by negroes with woolly hair. The other tribes of this continent seem to belong to a peculiar race, who rank in the lowest degree of civilization. Are then the negroes here likewise the aborigines? and have that wretched race yet been able to drive them before them into the extreme corner of their former country? Or have they come here later, and by sea? We do not know them as a sea-faring people.

We know scarcely any thing of the Haraforas, Alfuriers or Alfoirs, who have been confounded by many with the Papuas, from whom they however seem to differ. According to Leyden * they are among the most savage and oldest inhabitants of these islands, and are a peculiar race of men, with long hair, and often of a lighter colour than the Malays.

We do not find any ground in the historians of

* L. c. p. 217.

Manilla, for admitting on these islands a third race, different from the negroes and the civilized light-coloured inhabitants of the coasts.

The language of the Papuas, who live in insulated branches among other nations, without connection or community, must have been subdivided into many different dialects; the Malays of the peninsula of Malacca regard the dialects of the negro mountaineers as mere chirping, which can only be compared to the voice of larger birds; and, in many islands, no more favourable idea of it is entertained. The language of the Haraforas is considered as a peculiar one, which has nothing in common with the languages of the other tribes.* The Spaniards, on the contrary, affirm of the Aetas of the Philippines that in the rules of their idiom, there is a great coincidence with those of the inhabitants of the coasts, (Fra Juan de la Conception,) and that they speak dialects of the same language as the Indians (Zuñiaa.)

According to Forster, the languages of the tribes of his second race of men are not only totally different from the general language of the South-Sea islanders, but are even among themselves entirely distinct and dissimilar. The specimens given by him contain, however, besides the numerals, a few other roots, which are common to

* Leyden, l. c. p. 217 and 218. Marsden's Grammar, Introduction, p. 22.

them all, and the same observation may be applied to the vocabularies collected by Le Maire and Schouten in New Guinea and the Isle de Moise.

The languages in New Holland seem to differ among themselves, and from the dialects of the other race of men; the collections of words, which we possess of them, are, however, not sufficient for us to form an opinion. Sir Robert Brown assures us, that the tribes with whom he had any intercourse, could not count above four, and that five, and many, were with them synonymous.

We now come to the predominant race, of handsome physiognomy, long curly hair, and fair complexion, though more or less tanned by the influence of the climate; extending from Madagascar in the west, to Easter Island in the east.

We must acknowledge, with Marsden, the identity of the original stem to which all the dialects belong, that are spoken by so many different tribes, scattered over so immense a space. The coincidence of the numerals in all the dialects from Madagascar to Easter Island * proves, in strictness, only a common communication, and not the same descent. Numerals are easily taken from a foreign language; we find them the same in many dialects of the Papuas, of which the affinity is doubtful, and the inhabitants of the Marianes first

* See Hervas Arithmet. d. nat., and the Table of Comparison in Cook's third Voyage, Appendix, i.

forgot to count in their own language, when they accustomed themselves to the Spanish numbers.

We find in all dialects, besides similar numerals, a considerable number of common roots, which indicate, for the most part, simple things and ideas, and seem to be inherited from the same stock, but not learnt from a foreign people. We can point out these roots in the vocabularies of Madagascar, as well as in those of the islands of the Great Ocean.

Lastly, the grammar of the better known dialects of Malacca, Tagalog, and Tonga, is more or less formed, but in the main the same; and nothing authorises us to suppose that it is otherwise in those which are less known. The very simple system of the language, though the roots are polysyallabic, is nearly the same as in the monosyllabic languages. There is no declension; the roots stand either as in the Chinese, independent of each other, and receive their value from their place; or, in more cultivated dialects, are modified by particles differently annexed or interposed.

Many distinct tribes of this race of men, and speaking different languages, inhabit the islands of the East Indian archipelago. Leyden represents the pure dialect of the Javanese, as it is spoken in the interior of the island, as nearly and intimately related to the Sanscrit. The most simple objects and ideas are expressed by words, which seem to differ in the pronunciation alone from the Sanscrit, as the use of a less perfect alphabet necessarily implies.* The language, monuments, and history point back to India.

History first shows us, in the twelfth century, one of these tribes, the Malays, extending from the country of Manangkabau, in the south-west of Sumatra, their first abode, their conquests and the law of Mahomet, which they had received from Arab merchants, as well on the peninsula of Malacca, as on the coasts of the other islands. The converted nations were often confounded with them, and the expression Malays, Moors, and Mahometans employed as synonymous.

We find in the third book of Marco Polo, a picture of this archipelago at the end of the 13th century; and this picture is still like. The observations of this traveller, within the circuit of his own experience, are always faithful; and the fables he relates on the authority of others, are not yet forgotten in the places where he collected them. Pigafetta deserves similar praise. Marco Polo found that the people, who lived on the sea coast of the kingdom of Felech, in the island of Little Java, were Mahometans, who had received the law of Mahomet from merchants trading thither. Pigafetta, who was in Tidore in the year 1521, says, that about fifty years before, the Moors had conquered the Moluccas, and had in-

* Leyden, l. c. p. 190.

troduced their faith there. The vocabulary which he collected there, coincides with the present Malayan.

The Malayan has become the general language in this part of the world, the language of all trade and intercourse, and is spoken in the houses of the Europeans as far as the Cape of Good Hope. This language is perfectly well known to us; Marsden's Dictionary and Grammar, London, 1812, leave us nothing to wish for in this respect. The Introduction to the Grammar contains the history of the language, and the literature of the sources where it may be learnt.

The Malayan is a younger branch of the general stock. It contains, with a portion of common roots, a considerable number of Indian words, and Mahometanism has had a later but more superficial influence. The Arabian system of writing has superseded the Indian, to which, however, many heathen tribes still adhere. Only uninformed persons have considered as dialects, the four kinds of style in the general Malayan language, which are adapted to the rank and station of those by whom it is spoken; the language of the court, of the great people, the peasantry, and the market. In the Malay grammar we have a point of comparison for the other less known tongues of this branch.

We owe to the researches of the English our increasing knowledge of the people and languages of the East Indian Islands, and refer to the works already quoted; Marsden's Sumatra, Raffle's Java, the Asiatic Researches, the Asiatic Journal, &c. Their learning will succeed in decyphering the monuments of forgotten history in Java, in clearing up their manners and customs in connection with that of other people, in deriving the aboriginal people with whom we are engaged from Upper Asia, and in pointing out the way by which it has wandered to its present sea-encircled abodes.

The Philippines offer to us a peculiar family of the same people, and the same mother language. We here find the language in the highest perfection of its peculiar independent cultivation, and the elementary works of the different dialects, for which we are indebted to the Spanish missionaries, open to us a philological treasure, of which we shall endeavour to take a view.*

* Vocabulario de la lengua Tagala por el Padre Ivan de Noceda y el Padre Padro de San Lucas de la Comp. de Jesus. Impresso en Manila en la imprenta de la Comp. de Jesus, fol.

Vocabulario Tagalog por Fr. Pedro de Buenaventura, 1613.

Vocabulario de la lengua Tagala por nostro Hermano, Fr. Domingo de los Santos de Religiosos minores descalzos. Impressa en la muy noble villa de Tabayos, A.D. 1703, fol.

Idem, Reimpresso en la imprenta de N. S. de Loreto, Sampaloc, 1794.

Arte Tagalog por el Padre Fr. Francisco de San Joseph, 1610.

Arte de la lengua Tagala por el Padre Augusto de la Magdalena, 1669, 8vo.

Arte y reglas de la lengua Tagala, Thom. Ortiz, 4to.

Compendio de la Arte de la lengua Tagala por el Padre, Fr. Gaspar de San Augustin, Religioso de el mismo Orden, 1703.

The inhabitants of the coasts of these islands, whom we may consider as their first conquerors,

Idem, Secunda Impression en la imprenta de N. S. de Loreto, Sampaloc, 1787, 8vo.

Tagalismo elucidado y reducido (en lo possible) a la latinidad de Nebrija con su Syntaxis, tropos, prosodiæ, possiones etc. y con la alusion, que en su uso, y composition tiene con el Dialecto Chinico Mandarin, con las lenguas Hebrea y Griega. Por N. H. Fr. Melchor Oranguren de Santa Ynes, Religioso descalzo. Mexico en la imprenta de D. Fr. X. Sanchez, 1742, 4to.

Arte de la lengua Tagala y Manual Tagalog por Fr. Sebastian de Totanes de Religiosos decalzos de San Francisco. Impresso en la imprenta de N. S. de Loreto Sampaloc extra muros de la Ciudad de Manila, 1745, 4to.

Idem, Reimpresso en Sampaloc, 1796, 4to.

Bocabulario de Pampango por el muy R. P. Lector, Fr. Diego Bergaño de la Orden de los Hermitanos en Manila en el Conviente de N. S. de los Angeles, fol.

Arte de la lengua Pampanga por Fr. Diego Bergaño en la imprenta de la Comp. de Jesus, Manila, 1729, 4to.

Idem, Sampaloc, 1736, 4to.

Vocabulario de la lengua Bisaya compuesto por el R. P. Matheo Sanchez de la Comp. de Jesus, al Colegio de la S. C. de Jesus, Manila, 1711, fol.

Arte de la lengua Bisaya de la Provincia de Leyte, compuesta por el P. Domingo Ezguerra de la Comp. de Jesus. Tiene enxeridas algunas advertensias de la lengua de Zebu y Bool, 1662.

Idem, Reimpressa en Manila en la imprenta de la Comp. de Jesus, 1747, 4to.

Arte de la lengua Iloca por Fr. Lopez, Manila, 1617, 4to.

Vocabulario de las lenguas de Philippinas por Alonzo de Mentrida, 1637, 4to.

Arte de la lengua Bisaya y Bocabulario Español Bisaya de lengua Sugbuana compuesto por Fr. Thomas de San Geronimo de los descalzos de San Augustino. Reducido a ma exacto orden etc. por uno individuo de la misma Provincia. The manuscript is in our possession.

Los Indios of the Spaniards, speak, according to their tribes, seven different chief dialects; namely, in the north of Luçon the Pampangos, the Zambales, Pangasinanes, Ylocos, and Cayayanes, in the neighbourhood of Manilla the Tagalese, and on all the southern islands the Bisayas, with some idioms.*

The Spaniards are strangers in the Philippine islands. Many tribes of the Indians, even in the interior of Luçon, have maintained their independence, and those on the coasts, who, on being converted to Christianity, have submitted to a foreign yoke, have not learned the foreign language. The religious orders who accomplished the spiritual conquest of the country, and ensure the political supremacy, have adopted their language. The Tagalese, in particular, which, from the circumstance of its being spoken in the neighbourhood of the capital, is become the chief language, has been greatly enriched by them, not only with books to facilitate the acquisition of it, but also edifying compositions of all kinds in verse and prose. Fr. Francisco de San Joseph is called El Ciceron, and Fr. Pedro de Herrera, El Horacio Tagalo; there is even no want of tragedies. The *artes* and

* Natives of the Molucca islands were formerly settled in Marigondon, on the shores of the Great Bay of Manilla; their descendants, besides the Tagalese and the Spanish, still speak their own language, which they carefully preserve. F. Juan de la Conception, t. vii. p. 102.

vocabularies of the Pampango, Bisaya, and Yloco languages have been printed. The grammars of the other languages are MSS., and the copies of them are chiefly in the provinces, in the hands of the fathers.

According to the evidence of all the Tagalists, these seven dialects agree in essential particulars, both in the grammatical forms and in the roots. We ourselves have compared the grammars of the Tagala, Pampango, and Bisaya languages, and observed but very inconsiderable variations. If the difference in the pronunciation hinders the native of one province from immediately understanding that of another, yet a short time suffices to remove the difficulty, and he soon learns to recognize his own language. Consequently, what is said of the Tagalese is applicable to the other dialects.

Leyden in the Asiatic Researches, p. 207, has represented the Tagalese, Malay, Bugis, and Javanese as sister languages; he has reduced the more artificial construction of the Tagalese to the elements of the Malay, and proved in both the identity of the particles, upon which, in a system of language that has no declension whatever, all grammar depends.

Leyden appears to us not sufficiently to appreciate the meritorious diligence with which the Tagalists have reduced into a table of conjugations (which at least affords an easy view) the verb, variously combined with all the particles that modify it, with simple, double or half-double, and euphonically altered roots. It is indisputable, that with this advantage their representation of the Tagalese verb does not correspond with the original simplicity of the language, and strives to present one system where, in fact, another exists.

Articles and prepositions combined with the verb generally mark no more than a direct and indirect case. The plural, and not, as in the Malay, the singular, is particularly marked by a separate particle. The pronouns are the same as in the Malay, only more complete. Besides the two plurals of the first person (one of which includes, and the other excludes, the person addressed,*) there is also a dual of each of the three persons. The pronouns have, in the direct and indirect case, different forms. The root which expresses the action has particles prefixed, affixed, and inserted, which answer to the prepositions in our language, and mark in it the time, and the relations which we are used to express in the substantives and pronouns, either by declining them, or by accompanying prepositions; hence the three passives, to teach the meaning and use of which is the most difficult task of the Tagalists. We can place in a sentence only the subject or object of the action in the nominative, and mark the relation in the

* These two plurals of the first person, besides the present branch, are still found in the Quitchua, or Peruvian language.

verb itself. Active and passive, *amo et amor*, Danish, *Jeg elskir og elskes*. The Tagalists can place the subject, the object, the end, or the instrument, and the place of the action in the direct case, and express the relation by the verb. The sense decides what shall be distinguished and placed before as the nominative of the phrase, and the form of the verb conforms to it. Thus, in the phrase, "*Peter cut Malchus' ear off with the sword*," we can, at pleasure, lay the emphasis on *Peter*, (the subject) that which cuts, (active form), *the ear*, (the object) that which is cut, (first passive form with *y*), the *sword*, (the instrument), with which is cut, (second passive form with *in*,) and on *Malchus* (the place) on which is cut, (third passive form with *an*). The delicacy and difficulty of the language lie in the use. The same particles which modify the root as a verb, modify it also in similar combinations, as substantive and adjective. The word already compounded is treated as simple, farther compounded, riches produce riches, but there is no proper change of termination.

The Tagalese employ in their poetry verses, which, though proper to the language, put us in mind of the Spanish metre, by the number of the syllables, and a kind of rhyme or half-rhyme. They, however, have refused to adopt the more artificial canzone and sonnets which Father Francisco de San Joseph endeavoured to give them. We tried in vain to procure specimens of their original heathen songs, of which some are still in existence. Who regards in the country itself, the history, art, and antiquities of an oppressed people?

We give, from three different sources, the Tagalese alphabet, which is connected with the more ancient system of writing, of the people of the East Indian islands, and refer to the observations with which we accompanied it.

The inhabitants of the coasts of the island of Formosa, to the north of the Philippines, seem to us to belong to the same race, their language to the same original language.

We now come to the islands lying east of the Philippines, which we have considered as the first province of Polynesia. We find in their inhabitants, a family of nations, who are variously connected by the same manners and arts, by great skill in navigation and trade. They are a peaceable, amiable people, who worship no idols, live, without possessing domestic animals, on the gifts of the earth, and only offer to invisible gods the firstlings of the fruit on which they subsist. They build the most ingenious boats, and accomplish distant voyages, by means of great knowledge of the monsoons, the currents, and the stars. On the western islands, the Pelew islands, Eap, the Marianes, some customs of the East Indian islanders, such as the chewing of the betel, are found to have been introduced.

Notwithstanding the great resemblance of most of the tribes, and the frequent intercourse which unites them to each other, yet a great diversity of languages prevails among them.* We were called upon to collect specimens of their dialects, because we were more nearly connected with them than other scientific travellers who preceded us, and we give, in the appendix, a comparative list of words of the Marianas, Eap, Ulea, and Radack.

The people of the Marianas, according to Fra. Juan de la Conception, resembled the Bisayas, as well in appearance as in language, which latter differs, however, in several things, (*in algunas cosas alterado*). This Chamori or Mariana language has almost vanished with the people who spoke it, and the new generation speak the language of the conqueror, and their own only disfigured by a mixture with it. It is to be observed that they count only in Spanish, and it cost us trouble to procure the numerals of the Mariana language. On the other hand, it appears, that appellations from the Philippine languages have been given to many animals and objects introduced from foreign countries. Thus, in the Pelew islands, animals which have been introduced by

* Others, such as those of the Pelew islands, who are distinguished by their laxity of morals, and less knowledge of navigation, are probably strangers in the family.

the English, have received Malay names. (The goat, *Gaming*, Malay, *Kambing*.)

A Vocabulario de la Lengua Mariana, in the form of the vocabularies which we have of the languages of the Philippines, particularly of the Vocabulario Tagalog. by Fr. Domingo de los Santos, proceeding from the Jesuits, is still at Agaña; an *arte* seems to be wanting. This MS. is decaying unused, as the Spanish language suffices the present clergy for their office. We have endeavoured to investigate the grammatical construction of the Chamori language, and visited the fathers in Manilla, who presided over the missions of Guahon. Several had not learnt the language properly, and an old man was unable to give any account of it. The names of places in the Marianas, as in the Philippines, mostly end in *an*, a particle which, in the languages of the Philippines, marks the relation of place, and requires the third passive, and we find other signs of analogy, all of which are wanting in the dialects of the Carolina islands. Don Luis de Torres assured us, that in the Mariana and Ulea languages there is no declension. We must observe, that the words of the Mariana language, which we communicate for comparison, are not taken from the Vocabulario, for which we had no time, but written after our own orthography, from the pronunciation of Don Luis.

A vocabulary of the dialects spoken in the Pelew islands * is given by Wilson, and it were only to be wished, that the same diligence had been employed in clearing up the grammar, or that even a few specimens, some songs, had been communicated, which might give us an idea of it.

This work has more authority, in our opinion, than a slight collection of words hastily thrown together by a Spaniard, who furnished us with it in Manilla, and which we suppressed for this reason. It would only serve to show how differently the same sound may be understood and taken down by different nations.

We must apologize for the insufficiency of the collections of words of Eap, Ulea, and Radack, which we likewise communicate, without entering into the formation of the language. Let it be considered how suddenly and unexpectedly our friend and instructor Kadu left us. While these collections were forming, a means of acquiring information had been obtained by us, which gradually became more perfect, and the business of revising, correcting, and completing our work, of conversing upon abstract ideas, and treating upon the grammar, had been put off to a time which we were not destined to spend together.

* An Account of the Pelew Islands, from the Journals of Captain Henry Wilson, by George Keate, the fifth edition, London, 1803. Supplement, p. 63.

The natives of Radack, like the English, with a pronunciation difficult to be acquired, have not the faculty of easily understanding foreigners, nor of making themselves understood by them. We believe these dialects to be less simple in their construction than the dialect of Eastern Polynesia. We do not recognize, in different sentences, the roots which we expect to find in them, and the difficulty of mutual understanding seems to indicate this circumstance. The dialect of the Pelew islands seems to us to be the most different, and that of Radack to approach the nearest to the common language of the eastern South Sea islanders; and it was there we first found the arithmetical system, founded on the scale of twenty, as in New Zealand and the Sandwich islands, while the more western Carolinians, the Malays, and the Tagalese, use the pure decimal scale, which is also customary in Tonga.

Even within the limits assigned to these provinces, namely, in the south-west, nearest to the habitations of the Papuas, and to the Moluccas, we find several islands, the inhabitants of which were understood by natives of the Sandwich islands, and whose boats resembled those of Owhyee: we speak of Mavil's islands*; a circumstance which appears to us to deserve attention.

* Vide Arrowsmith's Chart of the Pacific Ocean, 1798, and Meare's Voy. p. 293.

It is well known that in New Zealand, in the islands of the second province, as far as Easter Island, which lies so remote to the east, and on the separate group of the Sandwich islands, there is only one people, which is almost every where at the same degree of civilization, has similar manners and customs, and speaks the same language, the dialects of which are distinguished only by local variations in the pronunciation; so that mariners who have collected words on one island, have often been able, with their help, to make themselves understood on others very remote; the natives of the Sandwich islands, by those of the Friendly islands; and Tupia, an islander of the latter group, by the New Zealanders.

We are obliged to Messrs. Mariner and T. Martin for a complete grammar of the dialect of Tonga *, which enables us to examine more particularly the language of Eastern Polynesia. We recognize in it the Malay system of language in the greatest simplicity, and, according to our opinion, in a state of undeveloped infancy. It is a pleasing, childish lisping, which can scarcely be called a language.

The Tonga language is more nearly allied to the infinitely more complicated Tagalese, than to the Malay; it has a more frequent use of the article, and particularly distinguishes the plurals

* An Account of the Natives of the Tonga Islands, from the Communications of M. W. Mariner, by T. Martin, M. D. London, 1818.

by particles. The pronouns are evidently the same, and it has, with the two plurals of the first person also a dual. The roots are employed without distinction for the substantive, adjective, or verb. In the verb the three tenses are expressed by separate particles (*adverbia*) as in the Malay. When two roots stand together, the first, as in other dialects, is the substantive, and the other the adjective.

Notwithstanding this simplicity, the dialect of Tonga, which is one of the most deviating, is perhaps also, at the same time, one of the most polished of Eastern Polynesia. Tonga lies on the western frontier nearest to the forelands, and the numeral system, as we have already observed, is not that of New Zealand and the Sandwich islands.

The language of the Sandwich islands really appeared to us still more childish than the dialect of Tonga does in its grammar. We found in it only two pronouns, *Wau* for the first person, *Hoe* for the second, and only two adverbs to determine the time of the action. *Mamure* for the future, *Mamoa* for the past time. The interrogative, or doubting particle, *Paha*, which is affixed, is very much used. *Nue*, and *Nue nue*, very and great, form the comparative and superlative. Some particles, mark, as prepositions, the relations of the substantives.*

* We cannot certainly assign the limits, of our own knowledge of the language of the Sandwich islands, as those of the language itself; but in other specimens of the languages of

The words formed, after the manner of children, by the repetition of one sound, in which the root has sometimes one meaning, sometimes another, and sometimes none at all, which are much more frequent in the common language of the eastern islands than in the western more cultivated dialects, in which, however, they are not wanting, give it a peculiar and pleasing character. *

The Owhyeeans have, besides, adopted many words from foreign nations with whom they trade, which, with their pronunciation, the want of some

---

Polynesia, and particularly in Nicholas' Voyage to New Zealand, London, 1817, we find no indication of a greater extent; and there, too, only two pronouns. Pronoun of the first person, Owhyee, *Wau;* New Zealand, *Aou;* Tonga, *Au,* perhaps the Tagalese *Aco;* Malay, *Ku.* Tonga has, besides, among others, *Gita;* Tagalese, *Quita;* Malay, *Kita.* Pronoun of the second person, Owhyee, *Hoe;* New Zealand, *Eakoe* or *Acquoi;* Tagalese, *Ycao;* Malay, *Ankau.* What confuses us the most in the study of these languages, is the different orthography, used by various collectors of words, and linguists. You must often be acquainted with the word to recognize it.

* *Moku-moku,* war. *Moku,* island, and European ship.
Make-make, to love, to like. *Make* or *mate,* to kill, to strike.
Mire-mire, to behold, to see.
Moe-moe and *moe,* to sleep.
Nome-nome, to speak, to say.
Hane-hane, to make.
Para-para, to draw.
Mi-mi, mingere,
Wite-wite, quick, sudden.
Rike-rike, just so, just the same.

letters, and the indifference of some others, are not easily to be recognized. The number of them daily increases, and they supersede the native words. *

The language of religion in the Sandwich islands is a peculiar one, different from the one now spoken, which the common man does not understand; it is probably the ancient uncorrupted language of the people, and ought to be one of the first objects of the scientific researches of the philosopher whom fate allows a longer stay in these islands. The accounts of Otaheite coincide † with this, and perhaps it was by means of this more ancient religious language that Tupia, who was one of the learned, understood the New Zealanders, since other common people of his nation did not succeed like him.

* The letters *r*, *l* and *n*, *k* and *t*, are of the same value. Examples of such words are, *kau-kau*, Chinese, *tschau-tschau*, for *païni*, to eat. *Pane-pane*, Chinese for *aïni*, *coïtus*, which foreign word seems to be euphemical, as, notwithstanding the general want of shame, decent Maolons avoid the other.

Pihi, English, fish, for *haiïna*, fish.

Neipa, English, knife.

Pikenene, Spanish *pequeno* for *käea*, small. We are surprised at finding the same word again, not only in New Zealand (Nicholas) but also among those published as Greenland words by Bernard O'Reilly, (Greenland, the adjacent Seas, and the North-west Passage,) London, 1818.

† We refer to the testimony of Mr. Marin, of whom we shall speak below.

It is well known that at Otaheite, on the accession of a new sovereign, or other similar occasions, words belonging to the general language are banished, and replaced by new ones. Such arbitrary changes have, in later times, caused the language of this island, which was formerly but little different from that of Owhyee, to depart more and more from it, so that the natives of the two islands do not now understand each other.

The following fact in the history of Owhyee, which we owe to a credible witness, a thinking and well-informed man, M. Marin, a Spaniard settled there, and which was confirmed to us by the natives, unexpectedly shows us this strange custom also in the Sandwich islands, and that in the most singular manner.

About the year 1800, Tamaahmaah, on occasion of the birth of a son, invented quite a new language, and began to introduce it. The newly-invented words were not related to any roots of the current language, nor derived from any of them; even the particles, which supply the grammatical forms, and are the connectives of the discourse, were transformed in a similar manner. It is said that some powerful chiefs, who were displeased at this metamorphosis, destroyed the child who had caused it, by poison. At his death the enterprise which had been undertaken at his birth was abolished. The old language was again adopted, and the new one forgotten. The innovation proceeded from Hana-rura to Owhyee, where Tamaahmaah at that time resided. M. Marin was at Owhyee, where it scarcely began to penetrate. When we asked M. Marin how such or such a word had been called in the new language, he consulted the natives of Hana-rura who were present, to all of whom the circumstance was well known, but who had mostly forgotten the new words. * M. Marin knew of no other instance of arbitrary change of language in these islands; Kadu· had acquired no idea of such a thing in the Caroline islands.

Man has emigrated from the great countries lying between Asia and New Holland, contrary to the course of the wind, from east to west, and has taken possession of all the points of land which rise from the Great Ocean, as far as the distant and detached island of Pascha, which lies far remote in the east. His language proves his origin. His manners, customs, and arts indicate it. His domestic animals, and useful plants, which have

* Thus we can give only an insufficient specimen of this entire creation of a language, which, though satisfactorily authenticated to us, so far exceeds the measure of our imagination that we do not venture to ask belief in it.

CURRENT LANGUAGE.	NEW LANGUAGE.
Kanaka	*Anna*, man.
Waheini	*Kararu*, woman.
Kokine	*Amid*, to walk.
Irio	*Japapa*, dog.

M. Marin pronounces it *Irio*, otherwise we hear *Lio*.

everywhere accompanied him, and which all belong to the Old World, tell us the coasts from which he brought them.*

We find the sugar-cane, the plantain, the paper mulberry tree, the *Hibiscus Populneus, Curcuma*, the calabash, the arum, yam roots, and sweet potatoes; lastly, among animals, the domestic fowl; in Easter Island, the bread-fruit tree, and other vegetable productions; the hog and dog as far as the Society, Marquesas, and Sandwich islands. The hog does not seem able to subsist on the low islands. New Zealand had only the dog, the Society islands only the hog, but the dog was there known by name, (*Ghuri* according to Forster, *Gooli* according to Mariner,) and we think we have found in the word *Giru*, in Radack, the same name, and a similar traditional knowledge of the same animal. The swine and the dog are wanting in all the islands of the first province.

Pigafetta first described in Tidor (Molucca

* It is undecided whether the hog and the dog were not found in Chili; and Humboldt has proved that the *Musa* (the plantain) was native in Mexico before the African was imported from the Canary islands (in the year 1516) to the West Indies. The bread-fruit tree and the paper-mulberry tree decisively belong to Eastern Asia exclusively, where alone those of the congenerous species are now found. The Indian sugar-cane was transplanted by the ancients into Sicily, and by us to America. Different kinds of *Arum*, *Dioscorea*, *Convolvulus*, and *Ipomœa* (taro, yams, and potatoes), are met with in both hemispheres, and demand a stricter examination, into which we have here no room to enter.

islands) the preparation of the stuff made of bast, usual in all the islands; and the same writer shows us the Bisayas of his time, with pierced and enlarged ear-laps, such as Forster found among the natives of Easter Island; a custom which the latter have already abandoned in our times, and which we still found prevalent in Radack and the Carolina islands.

It would probably be a vain attempt to trace back to one principle, and to one source, the sacred, diversely prohibitory laws and customs of the taboo, which separate the sexes, raise insuperable barriers between the classes of the people, and, differing in the different nations, are in all in the same spirit, the basis of social order, and to understand in their connection these human institutions, or to deduce them from the religious and civil system of other known nations. Here writing is wanted; and had we not the written document at hand, who would be able to discover, from the similar prohibitions and usages of the Jews, the mild spirit of the Mosaic legislation, which assigns even to the beasts a well measured right, and in which, besides, the idea of pure and impure, appears to be unfounded?* We are, besides, very far from taking it for granted, that every civil or religious

* We observe, by the way, without drawing any consequence from it, that the word *Taboo* occurs in the books of Moses, in the same sense as in the South Sea islands; a circumstance which has not been overlooked by the learned.

order has proceeded, as a complete whole, from one spirit. Such an edifice is often executed by history, which receives from the hand of chance the materials for its construction: and do we not even see the weak man turn back from a purely spiritual religion to polytheism, and place his vain earthly confidence on the material object, the block or the stone? Is it not easier to ourselves, as to other nations of the world, to believe in witchcraft, lying, and vain words, than to adhere to the spirit?

The disparity of classes, which is so deeply rooted among the islanders of the South Sea, the particular sanctity of some persons and families, who are independent of fortune and civil power, involuntarily remind us of India. The objection that the several casts of India are devoted to particular professions, modes of life, &c. is untenable. Such distinction cannot take place on these islands.

The voluntary death of the wife at the funeral of the husband in the Fije islands, and the similar custom in the family of the Tocitonga, in Tonga, likewise refers to India.*

If now the question is started, how, and at what time, an originally Asiatic people has dispersed itself over the most remote islands of the Great Ocean, contrary to the course of the wind, bringing with it its domestic animals and useful plants; how the different tribes, in their separation from

* Mariner's Tonga, vol. i. p. 330.

each other, could there preserve the same manners, and the same arts, and, notwithstanding the want of writing, which alone seems able to arrest the language in its tendency to change, and the habit of arbitrary innovations, yet speak only one common language, we can only confess our ignorance. The circumstances mentioned prove a simultaneous emigration from one point, and seem to point to a more modern epoch; but the infancy of the language, and, in many respects, of the people itself, seem to throw back the period into remote antiquity. Our first navigators found the people in the South Sea in the same state as they now are.

Monsoons and storms carried the seamen of the Carolinas as well to the west as to the east, and frequently as far as Radack, towards 180° of longitude from Greenwich. We can easily give an account of the peopling of this island. But we find, in this province, people speaking different languages, who are distinguished by more skill in navigation, and who possess no domestic animals. The name of the dog is known only in Radack, in the eastern dialects.* These people, with many resemblances in other respect, and perhaps a marked transition of languages, seem rather to divide the more easterly islands of the Great Ocean from the western countries, than to connect them.

* *Giru* and *Ghuri* cannot be derived with certainty from *Kuyuk* Malay, *Iro* Bisaya, *Aso* or *Ayam* Tagalese; *Irio* or *Lio* of the Sandwich islands approach nearer to the Bisaya.

The opinion of Zuñiga*, and of those who have attempted to deduce and explain the peopling of the islands of the Great Ocean, according to the course of the monsoon, from east to west, from America to Asia, is refuted.

In case it should appear, from examination, that sufficient grounds really existed, to recognize in the inhabitants of South America, and the islanders of the Great Ocean, or the people of Eastern Asia, the same aboriginal people, and in their languages the same branch, the inhabitants of the New World would, perhaps, rather, according to Molina's opinion, be derived from the Old World over the ocean, by the island chain of the second province, and against the course of the monsoon, or over New Zealand, and under the dominion of the varying winds.

We set aside the comparison which has been attempted to be made between the colossal statues of the island of Pascha, and the monuments of Peruvian architecture. We recognize in the former figures, which are made of a light volcanic stone, merely the common idol, which is to be found in the morai of most of the islands, and which are called in the Sandwich islands, *Akua*, gods; and in the Society islands, *Tighi*, spirits, souls.

We observed that the islands situated nearest to

* In the second chapter of the Historia de las Philipinas.

the American coast, the Galepagos, Juan Fernandez, &c. like all the scattered points of land in the Atlantic and Indian Ocean, lying far from the continent, were uninhabited: none of the Americans were sea-faring people.

Zuñiga hazards a conjecture that the language of the Araucanians and Patagonians must coincide, in the main, with the language of the Philippine islands, and without any means of investigation, builds upon this supposition. But this is not the case. †

We have found no coincidence between the roots of the Araucanian language and those of the original language which has occupied our attention. The numerals, the pronouns, are different. We might, perhaps, reduce the conjugation of the verb, and the declension of the substantive to the root, which always remains unchanged, and to which only particles are annexed, but these are always placed after; and in the manner, as well as in the sense of the composition, there is quite a peculiar spirit, that has nothing in common with the Malay and the Tagalese. The person is marked in the verb, namely,

* The Patagonians, the Puelci, or Puelchi, the Orientals, as the Araucanians call them, belong, as is well known, to the Chilian people, and speak the same language.

† We have consulted for the Araucanian language,

Bern. Havestadt, Chilidugu, Monast. 1777.

Molina Saggio sulla Storia Civile del Chili, Bologna, 1787.

Mithridates, iii. p. 403: and for the Quichua language, Mithridates, iii. p. 519.

at its termination; the personal terminations remain perfectly similar through all the tenses, and, in the essential points, through all the modes. By the insertion of different particles after the root, (only a few prepositions are placed before it,) a number of conjugations arise, the signification of which appears to be founded on different conditions. Thus negative, frequentative, &c. In the transitive conjugations (*Transiciones* of the Spanish grammarians) the object of the action, the pronoun of the accusative, is often taken into the verb. A sentence is often treated as the root of a verb, and provided with the particles of time, termination, person, &c. so that the sense is compressed into a single word. From verbs thus compounded, in the same manner as from simple ones, derivatives are formed by various terminations. The Araucanian has, in the declension and conjugation a dual, but it has not the double plural of the first person, which the Quichua language of Peru has in common with the languages of India. This coincidence in the Quichua is, however, purely accidental, and not founded on any internal relation. The Quichua is as foreign to the original language which has engaged us, as the Chilidugu, with which, notwithstanding a remarkable difference of the roots, it generally agrees in the grammar, and evidently belongs to the same system of language.

The perfect regularity of the Araucanian language, which, without any anomalies, follows the law as necessity, testifies a calm, undisturbed, independent development, with which no foreign mixture or influence has interfered. The termination *an*, which is frequently heard in the Araucanian language, and contributed to deceive Zuñiga, is quite distinct from the same termination in the Tagalese.

The languages, as well as the people, appeared to us entirely different; and we are therefore of opinion that they may justly be considered as distinct races. Common features unite the Araucanians with other American nations, in the same manner as they do the islanders of the Great Ocean with the other people of the East Indian islands, and with the difference of social order, manners, and customs, there remain only two points to be considered, which are certainly calculated to excite attention, and respecting which, not to change the view of the question, we will communicate what has been handed down to us.

The hog and the dog have names in the Araucanian language, whereas the other animals, introduced by the Spaniards, are distinguished by foreign names. According to Spanish orthography, the hog is called *Chancho*, after the Italian *Ciancio*, two different species of dogs *Quiltho* and *Thega ;* and Molina is inclined to believe that they were native previous to the invasion of the Spaniards, and brought by the aborogines, from the west over the sea.

Father Acosta, who wrote soon after the conquest, does not venture to decide whether the hog was originally in Peru, or whether it was brought by Europeans; we only observe that the names mentioned are quite alien to the languages of the South Sea and the East Indies.*

Burney, in his Chronological History of the Discoveries in the South Sea, v. iii. ch. 5. p. 187; mentions a place in *Hendrick Brouwer's Voyagie naar de Custen van Chili*, p. 72, where a beverage of the Chilians, near Valdivia, is spoken of, called *Cawau* also *Schitie*, and by others, with Italian orthography, *Cici*, which is prepared like the *Kava* or *Ava* of the South Sea, and only requires longer fermentation. The root from which it is made, is called *Inilie*. The drinking of the kava is a custom peculiar to the inhabitants of the eastern islands, which is totally unknown in the islands of the first province, as well as in the East Indian islands, though the plant itself grows there. We have collected the *Piper methysticum* in Guahon, and the very similar *Piper latifolium* in Luçon. It is not to be taken for granted that this noxious herb may grow in Chili, even if others supply its place; and we confess that this coincidence of the name is remarkable. We find no mention of this beverage in Molina.

* The hog is called in the Malay, *Babi*; Tagalese, *Babui*; in the languages of the South Sea, *Bua*, *Buacea*, *Buaha*, and *Pua*. For the names of the dog, vide a preceding note.

Burney, in the place quoted, seeks a resemblance between the Araucanian poncho, and the costume of the islanders of the Great Ocean, which we do not find; and we can lay no greater stress on an uncertain tradition of the Araucanians, according to which they came from the west, because another makes them come from the north, and a third represents them as having sprung out of the earth which they inhabit.

The result of our study, as well of history as of nature, is to represent man as very young on this old earth. In the strata of the mountains, the ruins of a more ancient world lie buried. The waters retire, animals and plants spread from different points, in various directions, over the surface of the earth: the mountains will separate the countries: man descends from his cradle, the high land of Asia, and, advancing on all sides, takes possession of the continent: he spreads in the west over Africa, where the sun colours the negro; and over Europe, where later immigrated tribes evidently speak in three-fold dialects the language of India.* The Papua, in the eastern countries, lying under the line, suffers, under similar influence, the same change as the African; or, perhaps belongs, with him, to one race. The Chinese

* In Europe only the Cantabrians and Celts can be called Autochtones, and only so far as their immigration and descent cannot be traced. The race of the Tschudi may be referred to other Asiatic races.

remains unchangeable in eastern Asia. Other branches spread themselves in the north of Asia; the north-east point of the Old World offers the road to the New: here the nations disperse, and become alien to each other; a certain resemblance allows us to assume a common race of men, but the languages have become entirely dissimilar. History shows us still in fresh remembrance an inundation of people, which pours itself from north to south, over the plain of Mexico, drives other tribes before it, leaving behind monuments of its passage, and faithfully preserves reminiscences of its original country, the high land of Asia.* Another branch, the Esquimaux, whose physiognomy shows the Mongol and Chinese race, issues from north Asia, over the northern edge of America to Greenland, preserves in both hemispheres the same language, the same mode of life, and the same arts. Lastly, a bold sea-fearing people, the Malay race, spreads from the S. E. point of Asia, over the habitations of the Papuas, as far as the most eastern and remotest islands of the Great Ocean; and the question is asked, whether, to the south of the line also, man opened himself a passage, by means of ships, from the Old to the New World?

We have a notion, that if a person provided with the necessary knowledge, could overlook and compare, all the languages spoken by man, he would

* Humboldt, Vues des Cordilleres, p. 152. &c.

recognize in them only different dialects, derived from one source, and would be able to trace back the roots and forms to one stock.

THE TAGALESE ALPHABET.

THE first of the subjoined alphabets is derived from the "Compendio de la Arte de la Lengua Tagala, por el Padre Fr. Gaspar de San Augustin. Segunda impression. Sampaloc. 1787."

The second, from the "Arte de la Lengua Bisaya, de la Provincia de Leyte, por el P. Domingo Ezguerra de la Comp. de Jesus. Reimpressa en Manila, 1747."

The third, from an "Arte de la Lengua Bisaya," MS. The Tagalists agree that their characters have been borrowed from the Malays. The Malays adopted, with Islamism, the Arabic characters, but the unlearned people of the interior of Sumatra and Java still use the alphabet founded on the principles of the Sanscrit or Deva Negri; and, according to Marsden *, written like the Sanscrit and European languages from left to right. Leyden contradicts this. According to him, the

* Grammar of the Malayan Language, by W. Marsden, London, 1812. 4to. p. 2.

Javanese alphabet is written from right to left, and the Batta alphabet, in Sumatra, from bottom to top, exactly contrary to the method of the Chinese. The Batta characters are carved on trees or sticks with a creese; the Lampung and Rajang are varieties of it, written on different materials in a different order. The Bugis in Celebes, seems to agree with the Javanese, in respect to the order in which it is written. *

We have not been able to procure, nor even to get a sight of any thing written or printed in the Tagalese characters; though this writing is not entirely out of use in some remote provinces, nobody in Manilla was able to give us any account of it, and the Tagalists leave us in doubt respecting the order in which it is written. †

In whatsoever order the above-mentioned alphabets are written, the Indian mode of writing is not to be mistaken in it. The Tagalese writing

* Asiatic Researches, vol. x. London edition, p. 158. On the Languages and Literature of the Indo-Chinese Nations, by T. Leyden, pp. 190. 193. 205.

† El modo de escribir era formando los renglones de alto abajo empexando por la isquierda, y acabando por la derecha al modo de los Hebreos y Chinos sus caracteres eran enteramente diversos de los nuestros, no tenian vocales, &c. Historia de Philippinas, por Fr. J. Martinez de Zuñiga. Sampaloc, 1803. p. 30. Solian antes de agora (y aun muchos oy dia) escrivir de abajo hazia arriba, poniendo el primer renglon hazia la mano izquierda. Ezguerra, l. c. p. 1. "They write on bamboo, palm, and plantain leaves." Poblacion de Manila.

seems to be, with respect to the vowels, the most simple and the most imperfect. *

* Vide the table of comparison of the characters of different nations, by C. W. Büttner, 2d edition. Göttingen, 1779, where the Tagalese occupies, on the five first tables, the 43d column; on the 6th, the 22d; and on the 7th, the 21st. The text which refers to it is wanting.

VOCABULARY.

OF THE DIALECTS CHAMORI, (MARIANA ISLANDS,) AND OF EAP, ULEA, AND RADACK.

Remark.

We have endeavoured to represent the sound by means of our German letters, as far as they sufficed. We have written a middle sound between A and O with Å; a very open *e* (the French *ai-j'aimais*) with A̋; a termination very nearly resembling the French nasal sound with *—ng:* instead of the German *w*, we have used the simple *v*, and borrowed out of the English alphabet the *w* and *th* for similar sounds. The *j* or *g* of the French,)†(of the Russian, occurs only in the word *Nagen.*

The accent lies mostly on the last syllable. To prevent the meeting, or accumulation of consonants, a vowel appears to be inserted for the sake of euphony.

It is hardly necessary to say, that none of those faults could be avoided, which are usually treated with indulgence in similar essays. Not to mention inevitable misunderstandings, our orthography is uncertain, as the pronunciation of our teacher, in languages not his own, was not to be depended upon. We heard at Radack, *Medid, Irud, Dilé,* which Kadu pronounced *Mesid, Irus, Thile.* We always doubted between *d, th,* and *s;* between *ch, k,* and *g,* &c. Of the last letters *ch,* or *k,* seemed to sound hard at the end of words, and, in conjunction, to incline to a softer *g, Inguch, Ingaga gamilate Rossa:* I understand not, I understand not the language Russia. *

* On mature consideration, it has been judged the most advisable to print the following vocabularies, precisely as they are in the German work, without attempting to reduce them to English orthography, as these words would, in all probability, be very different from what they would appear, if written by an Englishman, from the pronunciation of the natives themselves. We may add, that the long vowels in German have more resemblance to the French than to the English, the *u* however being pronounced like the English *oo.* The consonants have, in general, nearly the same power as in English; *sch* is invariably our *sh,* and not *sk,* as some pronounce it. *W* is pronounced like our *v,* the German *v* being pronounceed like *f.* It is for this reason, probably, that M. Chamisso uses the English *W,* as he calls it, our pronunciation of *w* being, in fact, foreign to the German language.— *Note of Translator.*

VOCABULARY

OF THE

DIALECTS CHAMORI (MARIANA ISLANDS), AND OF EAP, ULEA, AND RADACK.

NUMBERS.

TO BE COMPARED WITH AND TO COMPLETE THE TABLE IN COOK'S THIRD VOYAGE.

Tagalese.	*Pampango, after the Spanish Artes.*	*Bisaya.*	*Pelew Islands, after Wilson.*	*Item, after a Spanish MS.*
1. Ysa	Isa	Usa and Sayo	Tong	Dita
2. Dalva and Dalava	Adua	D,uhà	Oroo	Teru
3. Tatlô	Atlo	Tolo	Othey	Tedey
4. Apat	Apat	Upat	Oang	Oa
5. Limâ	Lima	Lima	Aeen	Oim
6. Anim	Anam	Unum	Malong	Malo
7. Pitó	Pitu	Pito	Oweth	Vis
8. Valò	Valo	Valo	Tei	Yay
9. Siyàm	Siam	Siam	Etew	Ytiu
10. Polô and Povô	Apulo	Polo	Mackoth	Magot.

	Chamori.	Numbers of a certain Measure of Length (Fathoms), in the same Language.	Eap.	Carolina Islands (Ulea), according to J. Wilson, in the Duff, 1797. After the German Orthography.	Ulea.	Radack.
1.	Hatjijai	Tac natjun	Rep	Eiota (Pota)	Eoth	Duon
2.	Huguijai	Tac hugua	Ru	Ruo	Rii	Ruo
3.	Totguijai	Tac tulum	Thalep	Tolu	Al	Dillu
4.	Fatfatái	Tac fatum	Eninger	Teia	Fahn	Emmen
5.	Limijai	Tac lima	Lahl	Lima	Lim	Lallim
6.	Gonmijai	Tac gonum	Nel	Honu	Ol	Dildinu
7.	Fedguijai	Tac guijai	Medelip	Feizu	Fis	Dildimemduon
8.	Gnalguijai	Tac gualum	Meruk	Warto	Oeil	Eidinu
9.	Siguijai	Tac siguam	Merep	Hivo	The-u	Eidinemduon
10.	Manutai	Tac manud	Ragach	Segga	Seik	Tjabudjet or Tjongaul

	Chamori.	Eap.	Ulea.	Radack.
11.			Seikamethéo	Tjabudjetmeduon
12.			Seikemeruo	Tjabudjetmeruo
13.			Seikemesilu	Tjabudjetmedillu
14.			Seikemefao	Tjabudjetmeemmen
15.			Seikemelimo	Tjabudjetmelallim
16.			Seikeméoco	Tjabudjetmedildinu
17.			Seikemefiso	Tjabudjetmedildinemduon

	Chamori.	Eap.	Ulea.	Radack.
18.			Seikemeolo	Tjabudjetmeeidinu
19.			Seikemetheuo	Tjabudjetmeidinemduon
20.	Huguanafulu	Repudegach	Rueg	Tjagoren.
30.	Tulungafulu	Thalepanath	Selig	Tjagorenmetjabudjet
40.	Fatfatnafulu	Eningenath	Faig	Ruagor
50.	Limangafulu	Lahlonath	Limeg	Ruagormetjabudjet
60.	Gonumnafulu	Nelonath	Oleg	Dillagor
70.	Fitinafulu	Medelipenath	Fisig	Dillagormetjabudjet
80.	Gualungafulu	Merugenath	Oalig	Eagor
90.	Siguanafulu	Merebenath	Théuég	Eagormetjabudjet
100.	Manud and Gatus	Raai	Semaul	Limmagor
120.				Dildinu
140.				Dildinemduon
160.				Eidinu
180.				Eidinumeduon
200.				Tjabugi
1000.	Tjalan	Wubiu	Theongoras	
		Tarep	Theotog	Duonot

Observation. In the arithmetical system of Radack, the scale is of 20, as in New Zealand and the eastern islands. The simple numbers go only to 5. 6 is formed of 3, 7 is 6 and 1, 8 is formed of 4, and 9 of 8 and 1. Tjabudjet is the usual 10. Tjongaul is said of men, ships, pandanus fruits, &c.

	Chamori.	*Eap.*	*Ulea.*	*Radack.*
The name. What is that called?	Nahan	Waoresingen	A̋tan	A̋tan
Exclamation of surprise		Eretam	Ilomaut	Irīo
———— of dissatisfaction		Wutävan	Tamaurel	Epada
I	Guaho	Igagk	Ngang	Nga
You	Hago			
Yes	Huu	Ier	Illa	Inga (and on the southern groups) Ja
No; also prohibition	Ahe	Matamat	Tabu or Buluath Ebin	Emo and Ap
There is none, it is wanting		Tari	Tor	Eitolok
God		Tàutup	Tautup	Jageach
The name of God		Engalap, the same at Ngoli, Mogemug, and Ulea. At Feis, Rongala: at Lamureck and Elath, Fuss: at Fojo, Lagé.		
Exclamation on religious offering			Wareganam gure Tautup!	Gidien Anis mne jeo
The people repeat			Tautup	Jeo!
The soul	Anti			
The man — men	Lahi	Pimohn	Måmoan	Måmoan
The body	Tatautau	Kainim	Kagel	Goen
The blood	Haga	Ratta	Ta	Wothagedig
The sweat (s. warm)		Áthu	Låss	Mnagaru
The head	Ulu	Elingeng	Methackitim	Emethackworra and Methackwarr

	Chamori.	*Eap.*	*Ulea.*	*Radack.*
The hair of the head	Gapunalu	Lalügel	Timui	Worra
Any thing like hair, fibres	Gapu			
Hair	Pulu			
The beard	Atschai, also the chin	Räp	Elsål	Koriak
The eyes	Mata, also the face	Eauteg	Matai	Medja
To see	Atan	Mutangarangai	Kolomethoa	Medimedi
The ears	Talanja	Ilig	Talengel	Talengel
To hear	Hungug	Gorungar	Erungerung	Rungerung
The nose	Guihin	Busemun	Wathel	Wathu
To smell		Foloboun	Easangi	Easangi
The mouth	Patjud	Langach	Eol	Langin
The teeth	Nifin	Mulech	Nir	Nir
The tongue	Hula	Athaen	Luel	Luel
The neck	Hagaga	Lügünag	Uel	Wuruwen
The breast	Hauf	Nüerungoren	Uwal	Ugel
The belly	Tudjan	Thugunem	Siel	Sien
The arm	Kanei	Pach	Bäï	Bän
The hand		Karovinarinepagh	Humutel	Laperinepeï
The fingers	Kalulud	Pugelipagh	Kasthel	Thanetheri
The thumb	Tamagath	Wagulinegah	Kasthelep	—
The leg	Adding	Ai	Petehl	Nen
The foot		Garovereven	Patepatelpetehl	Leporinen
The woman	Palauan	Wupin	Tabut, in Feis: Feivil	Gora and Redini
The breasts	Susu	Thithi	Thithi	Thithi
The milk	Tschugususu	Lengirén	Fåll	Fåll

	Chamori.	*Eap.*	*Ulea.*	*Radack.*
To suck	Pogsai			
Pregnant		Kaithien	Sasiel	Eliipesien, (see Great)
To bear, also to lay eggs		Korgoel	Sasiemelau	Emesalesal
The father	No word	Tamangen	Taman	Taman
The mother	No word	Langelin	Rehn	Rehn
The child		Vagk	Nagen	Nagen
The boy		Taraman	Taraman	Taraman
The girl		Wúlil	Tarvével	Lerrick
Twins (??)		Tathangen	Usi	
The eldest son		Ngani	Molles	Sän } Used only in Radack, among brothers and sisters—brother, sister.
The younger ones		Waïn	Usel	Sathen
The daughter		Olagen	Moengel	Inén
To adopt a child		Fagk (see Child)	Lá-eul	Nasi
The friends	Atjama, now mostly used only for lovers, as, the beloved.	Tafaveil	Marer	Sera
The old man		Pelewider	Malellap	Elallap
The young man		Waitiketihk	Oaétit	Enning
A chief	Tjamoro	Pilu Lamuniur, Kathegube and Meur, Ratulweli. Pelli, Ruwach (Rupack Wilson.)	Tamohn	Irud, or Irus. Tamohn already introduced. Besides this several denominations appear to indicate progression of ranks among the Irus.

	Chamori.	Eap.	Ulea.	Radack.	
One among the people		Tonepinau	Malegaffageu	Armesuan	
People, men				Loma	
Mishapen, natural deformity, a cripple		Botalip	Emmate	Ruwéwé	
A language, a word		Rewemaringach	Säckapatapat	Gamelat	
Do you understand?		Komenang	Kogela	Kosalage	
I understand		Kümenang	Űgüla	Űsala	
I do not understand		Thajonang	Ittagela	Ingach	
Talk, speak		Marangach	Kapatapat	Tattigalai	
Silence		Fahwach	Tangiel	Riap	
Crying	Agang	Taulul	Tataul	Lamuit	
Eating	Tjumatju	Thamunemun	Mogai	Mogai	He eats, drinks: Mogit
Drinking	Guminim	Thachu	Por	Bogai	Bogit
To drink a cocoa-nut				Gaga	Gagit
To invite to eat		Piwotuguai	Tattegalai	Gisäsirick	
To ask for something		Pigofanai	Kassiso	Läsoch	
To take	Tjuli	Mugol	Bulii, to Feis: Choli	Kabudri	
To give	Nahe	Areganam	Kalamuje	Kalamuje	
Buy, exchange	Fahan	Uaraifanam	Eamuje	Mojamuje	
I will not	Mungajo				
To be any where, to remain, to stay		Wairi	Emelega	Eberi	
To go	Humanau	Mahn	Galloch	Wailok	
To come	Mamaila	Meongrai	Maiga	Waidok	

	Chamori.	*Eap.*	*Ulea.*	*Radack.*
To fetch somebody, to call	Maila, come here! maila quini	Mahnemupinning	Vosangahsog	Gollali
Whither are you going?		Thingamanangan	Kowalaïa	Cathigit
There (to shew the way)	Adju			
To climb		Manangelang	Theusagk	Resach
To run	Malago	Mumill	Therr	Theser
To spring		Mooch	Ludt	Gäloch
To step	Mamockat			
To stumble, to fall in walking		Idol	Täparack	Ewong
To stand	Tumotughe	Tüling	Süsach	Süsach
To sit	Matatju	Permowut	Mathothi	Sithiet
To lie	Umassum	Mål (and sleep)	Űlloch	Wawu
To yawn		Pingesül	Mauloch	Mö
To stretch oneself		Dhadha	Reloch	
To sleep	Mahigu	Mål	Mädur	Mädur
To dream	Manguifi	Lickai	Thal	Thanack
To awake	Magmata			
To waken (*it.* to right an upset boat)	Pangun			
To laugh	Tschali	Minimin	Malikowot	Lea
To cry	Tangis	Thingejur	Kawasinng	Atang
He cries	Tumatangis			
To sneeze		Oingut	Mussi	Mussi
To cough		Tautol	Tagefach	Pogepoch

	Chamori.	*Eap.*	*Ulea.*	*Radack.*
To sigh		Poghovan	Nassetairack	Menuna
To be afraid		Kogethigau	Resumith	Ilubüch
To be ashamed		Ettamera	Emma	Essoch
To tremble	Laulau			
One that trembles	Laulaulau			
To itch	Makaka			
To scratch, also to rasp	Kassase	Gatål-gitigit	Moé and Ethat, at Feis: Rub	Irir
To tattoo, to draw		Kotau	Möck	Ao
Caressing by touching the noses		Farai	Feissong	Agomit, not customary among men at Radack.
To bend		Mogawornack	Kowaru	Gehli
To break		Mutar	Kopi	Kosai
To tear		Mukuruv	Katarra	Epeosach
To cut		Mithap	Kutovi	Mutemut
Good	Mauli	Jertam	Ilamoút	Eidara
Bad, of men	Abbale			
——, of things	Tailage			
Not much	Diddini	Thaneior	Teitolop	Ejet
Much	Meggai	Wéor	Etolop	Eor
Great	Dankulu	Poga	Eolep, at Feis: Mallilop	Elilp
Little	Dikiki	Wätich	Edigit, at Feis: Taraman.	Irick
High		Otoliang	Etageet	Etageet

	Chamori.	*Eap.*	*Ulea.*	*Radack.*
Low		Otawut	Ottatal	Ottatal
Above, over		Mungelang	Theusach	Resach
Below, under		Mulu	Theusi	Thuseni
Healthy		Kaitii	Sahtii	Edjaghu
Ill	Malango	Vaiamith	Emmedack	Emmedack
Right	Agapa	Wanegilei	Gilimera	Rear
Left	Akagui	Wanemetau	Giltschägil	Jeridili
Light		Wowaut	Eppel	Emmera
Heavy		Tomal	Ettau	Irro
Young	Paggun			
Old	Amku			
Fat		Bogha	Eculip, at Feis: Epalling	Eghasur
Slender		Poetiketik	Egetigeth	Egoirick
Dry		Mallick	Epellepell	Emora
Damp		Wogarda	Ollö	Eée
Cold		Ollüm	Isaleu	Päo
Warm, (see Sweat)		Eatho	Låss	Mnagaru
White		Umira	Ewuet	Emous
Black		Alit	Wol	Raran
Carmine		Eria	Lap	*wanting*
		Denominations for Colours are wanting.		
Cylindrical		Otapalo	Eulul	Eúlethilith
Four-cornered		Emetavan	Emetavan	Eurevan
Flat		Bogarathan	Etoilep	Erilep

	Chamori.	Eap.	Ulea.	Radack
Enough		Kaivel	Laimai	Emuit
Remote in time or extent, distant, old		Wutaurel and Taurel	Esaolog	Eddo
Here		Eroi	Iga	Idi
Now		Tharu	Igala	Gihn
Near		Utuwur	Egarep	Ebeágk
Look there (ecce)		Waram	Mathailai	Juéo
To scold, to be angry to pull the hair		Tabuel	Saṣegh	Emadirdir
To beat, to wound		Mliéau	Kauli	Mani
To kill	Punu	Mlieaungaim	Kauliwoimes	Manimaniimütch
To die	Matai	Kaim	Imütch	Imütch
Contest, War	Mumu	Matämal	Maul	Meidar
The dart, the lance	Gugudanun	Thillagk	Tilleg	Mari
The same not pointed, *it*; to blunt	Fudfud			
To throw		Mun	Kattevi	Kave
To hit		Ikan	Jel	Ellil
To miss		Theikan	Tarami	Tjapomele
The sling		Kaul	Kaul	Wuath
The double-headed lance		Tauwalach		Gilibilip
The drum		*wanting*	*wanting*	Adi
The beat of the drum when the enemy is still distant		*wanting*	*wanting*	Ringesipinen

	Chamori.	*Eap.*	*Ulea.*	*Radack.*
Item, battle	*wanting*	*wanting*	*wanting*	Pinneneme
Singing and dancing		Turu	Waruk	Eäp
A particular war dance		Walebong	Walebong	*wanting*
Another dance		Kapangach	Kapangach	*wanting*
The house	Guma	Naun	Ihm	Ihm
Driftwood		Eal	Kapepe	Gaimed
Driftwood with iron		Marauasai	Waleparang	Gaithoga
Iron	Lulu	Usai	Parang	Mäl
The head of an axe (a piece of iron hoop)		Usai	Parang	Mäl
The chisel (a nail, or a similar piece of iron)		Matai	Tété	Miré
The axe		Kol	Moil	Sisür
The knife (a sharpened muscle shell, and our knife)		Ear	Sar	Bogebog
	Guasaun	Tamathelai	Fasitte	Ragäloll
The whetstone (see stone)				
To whet	Guasa	Musuḿ	Taité	Timetim
Sharp	Malagdus			
To work as a carpenter		Mutoi	Falla	Ticketick
To sew		Munevit	Thigi	Dilledill
A wooden vessel, in the form of a ship		Thawi	Tapi	Tapi
A round ditto	Sahádjan			
To pound		Eoagil	Lovis	Komällis

	Chamori.	Eap.	Elea.	Radack.
A wreath		Iliau	Kabulipeu	Pellepel
A necklace		Maremar	Maremar	Maremar
Ornament for the ears		Tharau	Wot	Worr
Mat of pandanus leaves		War	Mang	Mang
The sleeping mat		War	Sagi	Sagi
A kind of stuff made of the fibres of the Banana plant		Waigi	Kou	*wanting*
The bast apron of the men				Mudirdir
The mat apron of the women				Thibidja
The man's dress		Thoú	Kapellepel	
The woman's dress		Platu and Jong	Kapellepel	
Curcuma powder		Rahn	Rahn	
Fishing line		Lam	Gau	Gäth
Fishing net		Teú	Uch	Kabiul
The boat, a ship	Sahadjan	Mu	Oa	Oa
The mast	Falina	Olian	Gkeus	Gisu
The sail	Laadjag	Lai	Ű ï	Usala
What carries the outrigger	Gahid			
The outrigger	Litja	Tham	Tham	Gubach
A rope		Tal	Tal	Tho

	Chamori.	*Eap.*	*Ulea.*	*Radack.*
The string			Kologol	Kologol
The ornamental string with which the apron is tied		Wukämu	Muril	Irick Tjabogon
The prow, and Stern of a boat under sail		Mitämu	Mol	Moan
To steer, rudder	Ulin	Bogailäat	Ekailioth	Djudjuve
To row, oar	Pogsai	Mamann	Fathell	Girgagi
To lose sight of land				
Driving with the current		Kaiau	Sasol	Esisäsalog
To wreck		Obogail	Eckail	Emarungerung
To turn		Mup	Thou	Ribadi
To tack	Lailai, and thence, on account of its going here and there.	Teltel	Tattagul	Arbuluul
Pieces of tortoise-shell, a kind of coin	Lailai			
Thin pieces of tortoise-shell on a string, another kind of coin	Alas			
Bathing and swimming of men	Numango	Monong	Evoloch	Aü
To dive	Lumuuf	Mulit	Esúlong	Esüloch
To rise again	Kahulu	Farangalang	Ewäsach	Oaloch

	Chamori.	*Eap.*	*Ulea.*	*Radack.*
The sun	Addau	Al	Al	Al
The moon (a month of thirty days)	Pulan	Pul	Moram	Alling
The stars	Putiun	Tuv	Fiss	Idiu
The Polar-star		Fissimogedigit	Fissimogedigit	Lemannemann
The shadow	Aninnig	Vahn	Eangal	Allil
The morning	Aggaan	Kairagan	Eral	Eral
The mid-day	Talluani	Kaimesu	Tajet	Tajet
The evening	Pupoeni	Kaiau	Thasuleal	Thülog
The night	Poeni	Kainep	Ebong	Ebong
A day	Haani			

Observations. The time in Radack, Ulea, and Eap, is calculated according to the number of nights and moons, (on the Marian Islands, by the day and moon. The Sandwich Islanders, also count by the nights Po). Ebony is likewise used in Radack for to-day. The word which we have taken for a year, (12 months) has remained very doubtful to us.

	Chamori.	*Eap.*	*Ulea.* (Don Luis de Torres.)	*Ulea.* (according to Kadu)	*Radack.*
A year		Wosu	Sewarak		Sewarak
The day before yesterday	Nigabrija		Talanginlallau		
Yesterday	Nigab	Fanop	Lallau	Lalo	Inné
To-day	Paagu		Ralai		Ebong
To-morrow	Agupa	Chabul	Lao	Walasu	Ildiu
The day after to-morrow	Agupanja	Langelat	Salangin	Watalangin	Tjagalat.
The third day			Watalangin		
The seventh day			Ranalal		

The Days of the Month on Ulea, according to Kadu.

The 1st Lingiling	The 7th Mesevel	The 13th Olomoal	The 19th Sopatemir	The 25th Ereve
- 2d Sigaur	- 8th Mesavol	- 14th Alat	- 20th Ortevalan	- 26th Eiï
- 3d Mesul	- 9th Mesadu	- 15th Ir	- 21st Olabugi	- 27th Erevi
- 4th Meseven	- 10th Tjabong	- 16th Ladi	- 22d Olahué	- 28th Euu
- 5th Neselim	- 11th Alabugi	- 17th Gilei	- 23d Olamahé	- 29th Evan
- 6th Mesaul	- 12th Oloboa	- 18th Kaira	- 24th Tamalaval	- 30th Etav

The twelve principal Points of the Compass in Ulea, according to Don Luis de Torres.

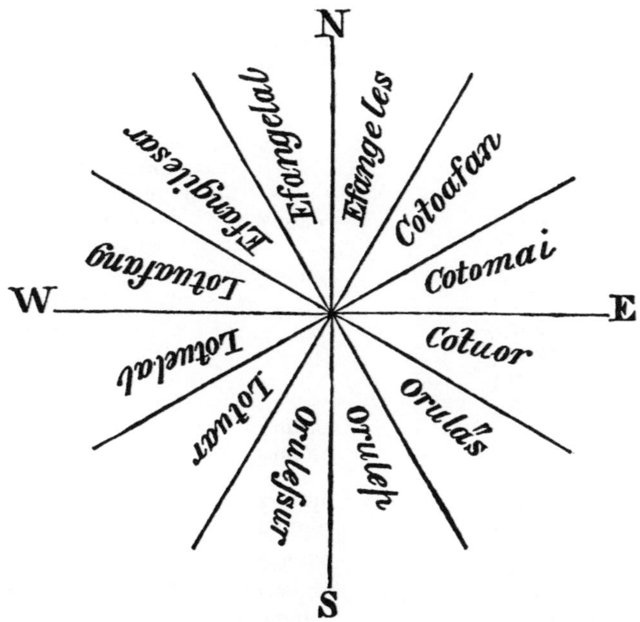

Half Rhumbs, being the numbers to 24. They are called by the following names, according to the two between which they lie:

Efangalab, caululor. Efangeles. Efangeles. caululor Coteafan, etc.

Course of the ships in Ulea according to the same.

Between W. and E. to the North, Puc.
Between E. and W. to South, Puilung.
Between N. and S. to East, Puitag.
Between S. and N. to the West, Puitug.

The Cardinal Points according to sun's place in the different times of the day at Ulea, from the same.

The Morning	Nissur.
The Noon	Egwol.
The Evening	Eppong.

	Chamori.	*Eap*	*Ulea.*	*Radack.*
North	Timi	Laélot	Maévan	Wasogien
South	Seplun	Imut	Majürr	Wasogrick
East	Manuu	Ngaäck	Mattaral	Kåsu
West	Faniipan	Ngal	Mäleso	Kåsusogepiling
The sky	Langin	Lang	Lang	Lang
The wind	Mangeu	Niveng	Aang	Githu
The sea	Tahsi	Näo	Lao	No
The flood		Fasach	Fasach	Äathagk
The ebb		Eisowil	Eisowil	Äatho
The current		Eatsch	Eath	Äathagk
The breakers		Tanna	Faleram	Thiwanegelinig
To throw out		Kaipaht	Sapat	Eotheck
Land, an island, a territory in an island	Tahno	Wunau	Valli	Enni
A low island group	Tahno	Lügülling	Lügülleng	Aäleng
The interior; the situation in the inner sea				Iar
The exterior; the situation on the outer sea		Illüch	Illüch	Iligieth
A passage		Thibutol	Thau	Tjer
A mountain	Alumtano	Taït	*wanting*	*wanting*
A river	Saddug	Lull	*wanting*	*wanting*
A water pit		Rahn	Tahl	Ranlibul
Sweet water	Hanum	Munum	Eliimi	Ira
Fire	Guafi	Nevi	Eaf	Gidieg

	Chamori.	*Eap.*	*Ulea.*	*Radack.*
To kindle				
Smoke	Assu	Muruweg	Fissigi	Dilé and Thilé
Clouds	Mapagahis	Athanenevi	Oath	Oath
Fog		Tharami	Tharami	Tharami
Rain.	Utjan	Thap	Thap	Thap
The rainbow	Issa	Nu	Uth	Uth
The thunder	Hulu	Laulüor	Laulüor	Tamäluth
Lightning	Lammlamm (i. e. shining.)			
Earthquake	Linau	Hirru	Fallamar	*wanting*
A path, way	Tialan	Ua	Ieal	Ial
A stone	Atju	Malang	Vas	Ragha
Fall, said of things		Emul	Eponloch	Ewonloch
A herb, tree, shrub, also forest		Pan	Oluel	Mar
A tree	Uddunhadju			
The trunk of a tree, wood	Hadju			
The root	Hali	Likangèn	Oagar	Oagar
The leaf	Hagun	Imm or Iuan	Teúl	Pellepel
The flower	Tschinali	Oamangin	Ual	Länn
To plant or sow		Miong	Fasagü	Gallub
To dig up the root	Haali			
Work			Engang	Mariliir
An herb	Papaguan			

	Chamori.	*Eap.*	*Ulea.*	*Radack.*
Th pandanus and its euit		Ner	Faht	Wob, the masculine tree; Digar, the wild; Eruan, cultivated varieties: Buger, Bugien, Eilugk, Undaim, Erugk, Lerro, Adiburik, Eideboton, Eremamugk, Tabenebogk, Radilebil, Tumulisien, Lugulugubelan, Aodian, Ulidien, &c. The sweet meat Moghan prepared from the juice.
The bast fibres of the cocoa nut, also to clean the nut from them.		Thaivu	Pajöl	Aé
Cocoa, the tree and the nut.	Nidju	Niu	Ni	Ni
The bread fruit tree and fruit	Lemmai	Ethau	Mä	Mä
The Sanane	Tjodha	Pao	Ut	Kaibaran
Arum esculentum		Mal	Eoth	Kadack
Arum sagittifolium		Ulack	Wulach	Jerat
Arum macrorhizon		Lai	Villa	Woth
Tacca pinnatifida	A place at Guahon is called Mungemung	Mogemug	Mogemug	Mogemug
Hibiscus populneus		Gahl	Giliven	Lo

	Chamori.	*Eap.*	*Ulea.*	*Radack.*
A branch, with useful bast		Aromä	Aromä	Aromä
A plant Triumfetta procumbens		Korach	Karach	Alahat
Curcuma		Gutol	Eong	*wanting*
Bambus		Mor	Wowau	*wanting*
Areca catechu		Bu	*wanting*	*wanting*
Caryophyllus aromaticus		Tongach	*wanting*	*wanting*
A kind of sweet potatoe		Rämot	*wanting*	*wanting*
Our quadrupeds				Giru, to compare with Gh-uri, the day in New Zealand, and in the Friendly Islands.

Nota, in the Pelew Islands, Oxen are called *Ming*, Goats, *Gaming*.

	Chamori.	*Eap.*	*Ulea.*	*Radack.*
Hogs	Babui, as in the Tagalese			
Cats (Spanish)		Gato	Gato	
Rats		Warro	Git	Gidirik
The train or tail of an animal		Wuck	Patal	Logon
The sea turtle		Woel	Woal	Uen
Great lizards	Iguana	Kaluv	*wanting*	*wanting*
Lizards		Athavaruru	Purupur	Uioe
A fish	Guihan	Nich	Igk	Igk
To swim	Numango	Kajen	Illoch	Illoch
A dolphin		Gätich	Gui	Gui
A shark		Kojong	Paghu	Paghu

	Chamori.	Eap.	Ulea.	Radack.
Ray (resembling Raga Pastinaca, or Raga Aquila)				Samuso Thotho
The flying fish		Kogk	Mongar	
Triton's horn shell		Babul	Taui	Silimaré, a general denomination for shell is wanting.
Sea hedgehog			Mugol	Mugol
A bird	Gaga	Eretä	Girigagk	Waó
To fly	Gumupu	Gaitomgagk	Elsoch	Gāsach
Feather		Fath	Ụlellemell	Emmerim
Nest		Taggil	Ḟa	Rong
Eggs		Fagk	Fathiel	Lip
To hatch		Bernasakein	Eponfathiel	Wavulerong
The cock		Nümen	Mallich	Kahu
The hen		Nümenewupin	Malugofeivil	Lala
The frigate		Molov	Gutaf	Agk
An ant				Kallep

SONGS OF RADUCK.

1.

(Sung by Women.)

:,: Esülog o no logo dildinu
Oalog o no logo dildinu :,:
dildinemduon!

:,: Dive in the sea six times,
Rise from the sea six times, (repeated six times)
Seven times!

2.

Wongusagelig, the Chief of Ligiep, headed his boats and people, to join Lamaray at Aur, when those of Meduro and Arno made war upon him. The first part of the Song represents his departure from Ligiep; the second, his arrival at Aur.

Wónguságelig
:,: Agaratérageríg
Iligiéth a lomá
„Wágeság diwón.
„Ríbadi aälénginé!
Esisäsalóg!
Aäthó! Aäthó!
Wónguságelíg! :,:
Eaaínewarasach:
„ Sellesi inneseo!
„Eyeweapwesog
„Tjabogén djudjuvé! djudjuvé! djudjuvé!
djudjuvé! djudjuvé! djudjuvé!
„Emarúngerúng aäthagín!"

Wongusagelig
Goes under sail;
On the beach the people throng!
Shift the sails round!
Strike we not on the reef!
Land out of sight!
Ebb! Ebb!
Wongusagelig, (repeated.)
And there resounds the command,
Keep the ships together!
There dashes the wave surely in!
To the ship before steer! steer! steer!
steer! steer! steer!
Carries away us, the flood!"

END OF THE SECOND VOLUME.

London:
Printed for A. & R. Spottiswoode,
New-Street-Square.

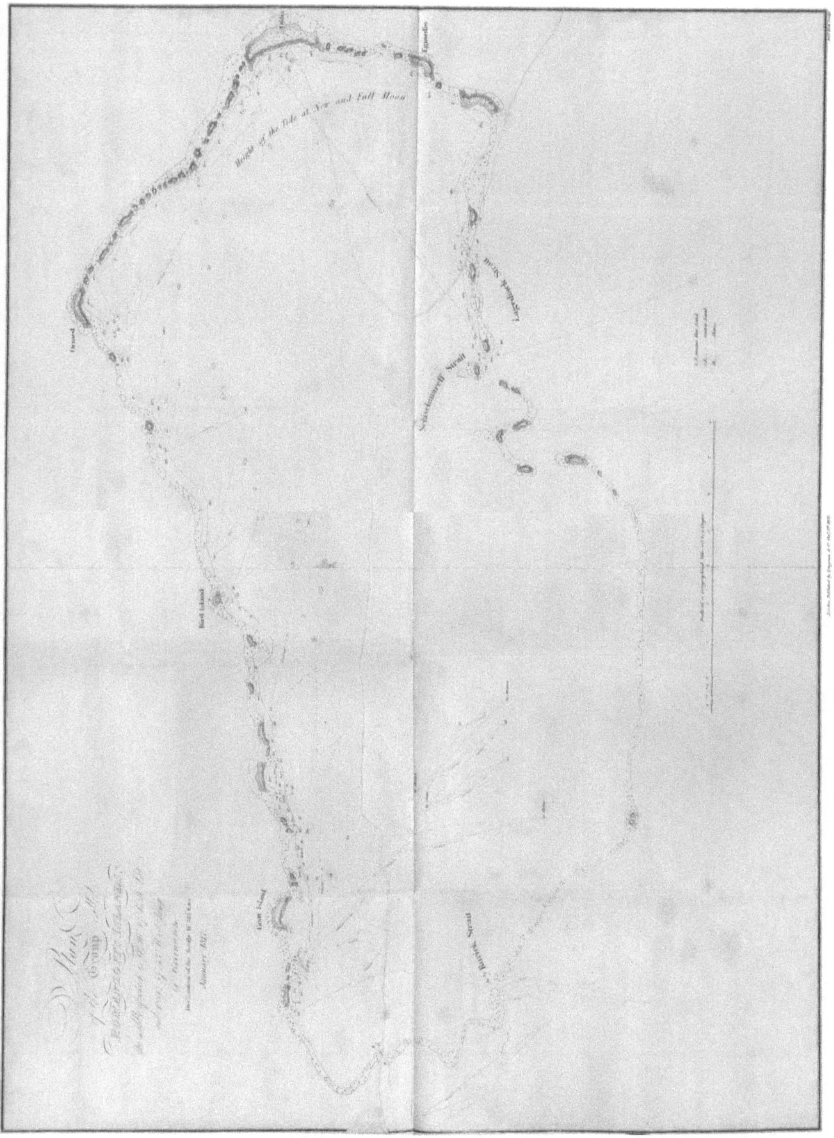

The material originally positioned here is too large for reproduction in this reissue. A PDF can be downloaded from the web address given on page iv of this book, by clicking on 'Resources Available'.

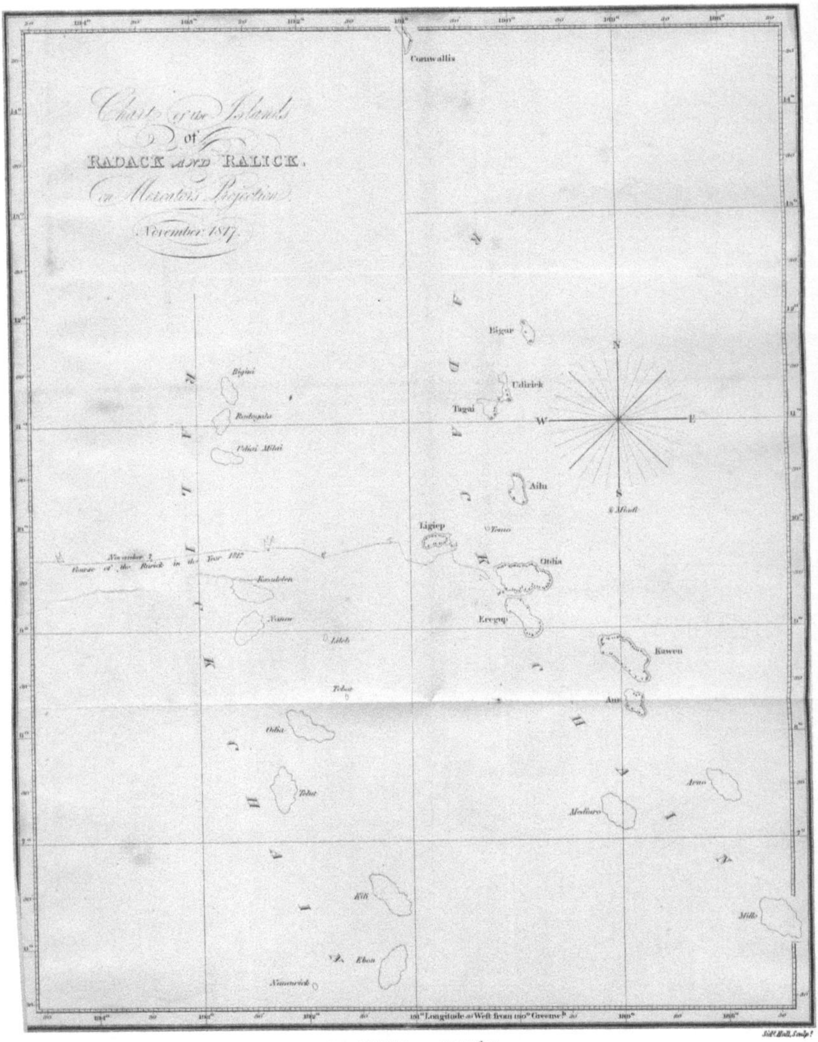

London, Published by Longman & Co. Octr. 8th 1821.

Sidy. Hall, Sculpt.

The material originally positioned here is too large for reproduction in this reissue. A PDF can be downloaded from the web address given on page iv of this book, by clicking on 'Resources Available'.

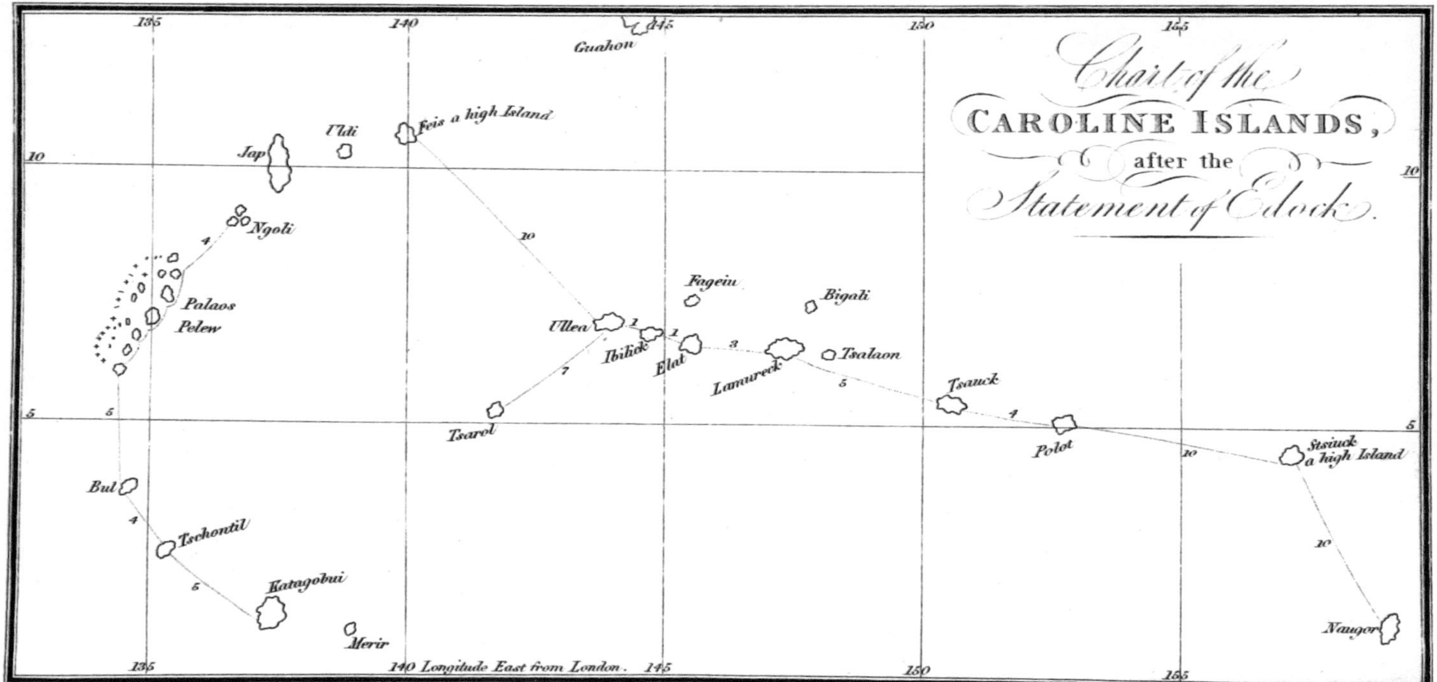

London; Published by Longman & C.o Oct.r 8th 1821.

Sid.y Hall, Sculp.t

For EU product safety concerns, contact us at Calle de José Abascal, 56–1°,
28003 Madrid, Spain or eugpsr@cambridge.org.

www.ingramcontent.com/pod-product-compliance
Ingram Content Group UK Ltd.
Pitfield, Milton Keynes, MK11 3LW, UK
UKHW041140300926
15471UKWH00011B/14
9781108057585